Primer

Instruction Manual

By Steven P. Demme

Math·U·See.

1-888-854-MATH (6284) - www.MathUSee.com
Sales@MathUSee.com

Primer Instruction Manual: Introduction to Math
©2012 Math-U-See, Inc.
Published and distributed by Demme Learning

www.MathUSee.com

1-888-854-6284 or +1 717-283-1448 | www.demmelearning.com
Lancaster, Pennsylvania USA

ISBN 978-1-60826-078-2
Revision Code 0616

Printed in the United States of America by Bindery Associates LLC

For information regarding CPSIA on this printed material call: 1-888-854-6284
and provide reference #0616-07272016

Building Understanding in Teachers and Students to Nurture a Lifelong Love of Learning

At Math-U-See, our goal is to build understanding for all students.

We believe that education should be relevant, skills-based, and built on previous learning. Because students have a variety of learning styles, we believe education should be multi-sensory. While some memorization is necessary to learn math facts and formulas, students also must be able to apply this knowledge in real-life situations.

Math-U-See is proud to partner with teachers and parents as we use these principles of education to **build lifelong learners.**

Curriculum Sequence

$\int$	**Calculus**
$\cos$	**PreCalculus** with Trigonometry
xy	**Algebra 2**
Δ	**Geometry**
x^2	**Algebra 1**
x	**Pre-Algebra**
ζ	**Zeta** Decimals and Percents
ε	**Epsilon** Fractions
δ	**Delta** Division
γ	**Gamma** Multiplication
β	**Beta** Multiple-Digit Addition and Subtraction
α	**Alpha** Single-Digit Addition and Subtraction
P	**Primer** Introducing Math

Math-U-See is a complete, K-12 math curriculum that uses manipulatives to illustrate and teach math concepts. We strive toward "Building Understanding" by using a mastery-based approach suitable for all levels and learning preferences. While each book concentrates on a specific theme, other math topics are introduced where appropriate. Subsequent books continuously review and integrate topics and concepts presented in previous levels.

Where to Start

Because Math-U-See is mastery-based, students may start at any level. We use the Greek alphabet to show the sequence of concepts taught rather than the grade level. Go to MathUSee.com for more placement help.

Each level builds on previously learned skills to prepare a solid foundation so the student is then ready to apply these concepts to algebra and other upper-level courses.

Major concepts and skills for Primer:

- Counting objects and developing numeracy
- Understanding place value
- Recognizing number names and symbols
- Reading and writing numerals
- Understanding addition and subtraction with concrete and representational models
- Understanding and writing the symbols +, −, and =
- Addition and subtraction of select numbers

Additional concepts and skills for Primer:

- Telling and writing time by hours and minutes
- Classifying squares, rectangles, triangles, and circles
- Reading and writing tally marks
- Recognizing patterns and finding differences
- Counting by 2s, 5s, and 10s
- Comparing and describing numbers and objects

Find more information and products at MathUSee.com

Contents

Curriculum Sequence 4

How To Use 7

Lesson 01   **Number Recognition** 11

Lesson 02   **Writing Numerals** 15

Lesson 03   **Number Recognition and Writing Numerals** 19

Lesson 04   **Geometric Shapes: Rectangles** 23

Lesson 05   **Number Recognition and Writing Numerals** 25

Lesson 06   **Geometric Shapes: Circles** 27

Lesson 07   **Number Recognition and Writing Numerals** 29

Lesson 08   **Geometric Shapes: Triangles** 31

Lesson 09   **Place Value: Units and Tens** 33

Lesson 10   **Place Value: Hundreds** 37

Lesson 11   **Unit Bars** 41

Lesson 12   **Addition: Introduction and Symbol** 45

Lesson 13   **Addition: +1** 47

Lesson 14   **Counting to 20** 49

Lesson 15   **Addition: 2 + 2 and 3 + 3; Vertical Addition** 53

Lesson 16   **Shapes: Squares; Addition: 4 + 4 and 5 + 5** 57

Lesson 17   **Skip Counting by Two** 59

Lesson 18   **Addition of Tens** 63

Lesson 19   **Skip Counting by 10** 65

Lesson 20   **Addition of Hundreds** 67

Lesson 21   **Solving for an Unknown** 69

Lesson 22   **Skip Counting by Five** 73

Lesson 23   **Tally Marks** 75

Lesson 24   **Addition: Making 10** 77

Lesson 25   **Skip Count to Find Area** 81

Lesson 26   **Telling Time with Minutes** 83

Lesson 27   **Telling Time with Hours** 85

Lesson 28   **Telling Time with Minutes and Hours** 87

Lesson 29   **Subtraction: Introduction and Symbol** 89

Lesson 30   **Subtraction: −1** 93

Student Solutions 95

Master Index for General Math 125

Primer Index 127

HOW TO USE

Five Minutes for Success

Welcome to *Primer*. I believe you will have a positive experience with the unique Math-U-See approach to teaching math. These first few pages explain the essence of the methodology, which has worked for thousands of students and teachers. I hope you will take five minutes and read through these steps carefully.

If you are using the program properly and still need additional help, you may visit us online at MathUSee.com or call us at 888-854-6284.

–**Steve Demme**

Special Instructions for *Primer*

In *Primer* we are giving students a taste of several math concepts but not requiring them to master these concepts, so make it fun. Use the blocks as much as possible. Let the students develop a positive attitude towards numbers and number concepts as they are introduced to math for the first time. Attitudes are catching. As you and your student build with the blocks, play the games in the lessons, and enjoy the subject, it will set a positive tone for the class. The concepts taught in *Primer* will be introduced again and taught for mastery beginning in *Alpha*.

Support and Resources

Math-U-See has a number of resources to help you in the educational process.

Many of our customer service representatives have been with us for over 10 years. They are able to answer your questions, help you place your student in the appropriate level, and provide knowledgeable support throughout the school year.

Visit MathUSee.com to use our many online resources, find out when we will be in your neighborhood, and connect with us on social media.

The Goal of Math-U-See

The underlying assumption or premise of Math-U-See is that the reason we study math is to apply math in everyday situations. Our goal is to help produce confident problem solvers who enjoy the study of math. These are students who learn their math facts, rules, and formulas and are able to use this knowledge to solve word problems and real-life applications. Therefore, the study of math is much more than simply committing to memory a list of facts. It includes memorization, but it also encompasses learning the underlying concepts of math that are critical to successful problem solving.

The Suggested 3-Step Math-U-See Approach

In order to train students to be confident problem solvers, here are the three steps that I suggest you use to get the most from Math-U-See *Primer*.

Step 1. Prepare for the lesson
Step 2. Present and explore the new concept together
Step 3. Practice for the student

Step 1. Prepare for the lesson

Watch the video lesson to learn the new concept and see how to demonstrate this concept with the manipulatives when applicable. Study the written explanations and examples in the instruction manual.

Step 2. Present and explore the new concept together

Present the new concept to your student. Have the student watch the video lesson with you, if you think it would be helpful. The following method should happen interactively.

a. **Build:** Use the manipulatives to demonstrate and model problems from the instruction manual. If you need more examples, use the appropriate lesson practice pages.

b. **Write:** Show the problems on paper as you build them with the blocks.

c. **Say:** Talk through the math concept as you build and write.

Young children think differently than adults do. They need to see, touch, and build with concrete objects in order to understand. Movement is an important part of learning for young children because they take in information through their senses and their muscles. Use "Build, Write, Say" as long as necessary for the student to be comfortable with the new material. One of the joys of teaching is hearing a student say, *"Now I get it!"* or *"Now I see it!"*

Step 3. Practice for the Student

Using the lesson practice problems from the student workbook, have students practice the new concept. Coach them through the building, writing, and saying process. Work together to do as many pages as necessary for the child to be comfortable with the concept before moving to independent work. Do not be concerned if the student does not master a concept in *Primer*. All of the concepts introduced at this level will be taught again in future levels of Math-U-See. Use the application and enrichment pages for a variety of activities that reinforce the skills taught in *Primer*. You can also go to MathUSee.com to find other resources.

Length of a Lesson

How long should a lesson take? This will vary from student to student and from topic to topic. You may spend a day on a new topic, or you may spend several days. There are so many factors that influence this process that it is impossible to predict the length of time from one lesson to another. If you move from lesson to lesson too quickly, the student will become overwhelmed and discouraged. If you move too slowly, your student may become bored and lose interest in math. I believe that as you regularly spend time working along with your student, you will sense when to proceed to the next topic.

Tell me, I forget. Show me, I understand. Let me do it, I remember.
                                                                    –Ancient Proverb

Number Recognition
Counting from 0–9

The two skills needed to function in the decimal system are the ability to count to nine and an understanding of *place value*. In the *decimal system*, where everything is based on 10 (deci), you count to nine and then start over. To illustrate this, count the following numbers slowly: 800, 900, 1,000. We read these as eight hundred, nine hundred, one thousand. Now read these: 80, 90, 100. These are read as eighty, ninety, one hundred. Once you can count, work on place value. The two keys are learning the counting numbers zero through nine, which tell us how many, and understanding place value, which tells us what kind.

When counting, we begin with zero and then proceed to nine. Traditionally we've started with one and counted to 10. Look at the two charts that follow and see which is more logical.

1	2	3	4	5	6	7	8	9	10		0	1	2	3	4	5	6	7	8	9
11	12	13	14	15	16	17	18	19	20		10	11	12	13	14	15	16	17	18	19
21	22	23	24	25	26	27	28	29	30		20	21	22	23	24	25	26	27	28	29

The second chart has all single-digit numbers in the first line, and then, in the second line, each digit in the units place is preceded by a one in the tens place. In the third line, each digit in the units place is preceded by two in the tens place. The first chart, though looking more familiar, has the 10, the 20, and the 30 in the wrong lines. When counting, begin with zero, count to nine, and then start over.

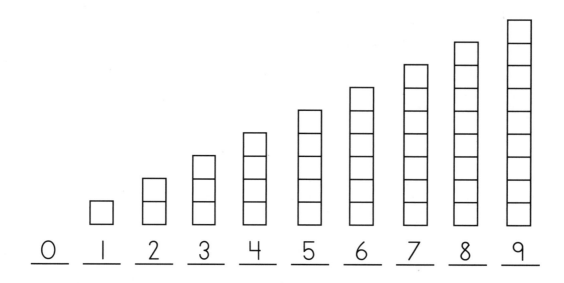

On the worksheets, the squares are the same size as the green unit blocks, so you can place the bars on the paper, count them, and then circle the correct number. When there are objects and pictures, count those and then circle the correct number. You can also have students trace over the numerals they circled to prepare them for writing. Remember to begin with zero when counting. Notice the senses involved: see the number of squares (visual), hear the correct number as you count (auditory), and build with the blocks (kinesthetic and tactile). Another way to combine the senses is to give a number verbally and have the student clap or jump that many times.

One of the reasons we use blocks is so that students recognize that numbers represent real things. On the worksheets, you will notice that we move from blocks to counting pictures of things, like frogs, crayons, and shoes. When counting real objects, be sure to arrange them in different patterns. For example, arrange the unit blocks in a circle as well as in a line.

Example 1

Example 2

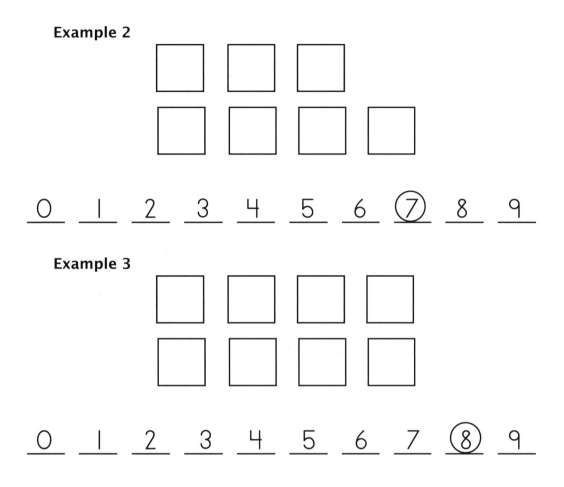

0 | 2   3   4   5   6   ⑦   8   9

Example 3

0 | 2   3   4   5   6   7   ⑧   9

There are solution pages at the back of this instruction manual. The solutions for pages A and D are shown for each lesson.

Writing Numerals

These exercises have two answer choices. The student is given the number of blocks and must choose the correct numeral that corresponds to the number of blocks. Take the green unit blocks and place them on the squares in the book. Then circle the correct answer and cross out the incorrect answer. Trace the correct answer and say it, as shown in Examples 1 and 2.

Example 1

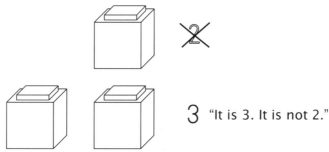

3 "It is 3. It is not 2."

Example 2

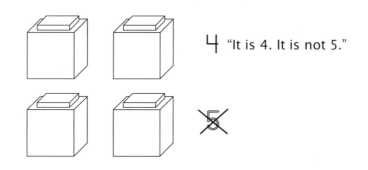

4 "It is 4. It is not 5."

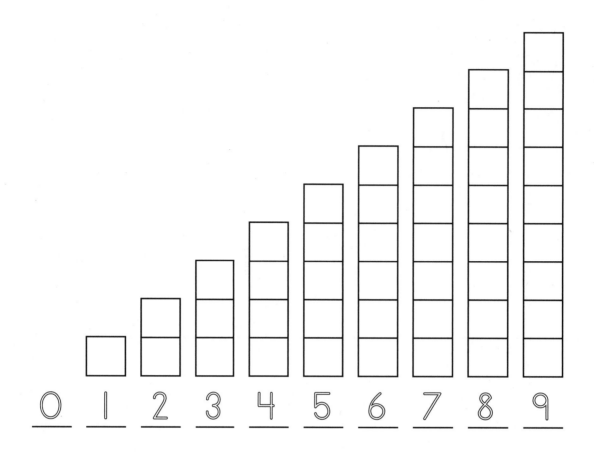

You may choose different ways to have your student write the numerals. Please teach your student that there is more than one way to write the numeral "4." It may have a closed top, as shown in many books, or it may be open at the top, as we have written it on the worksheets.

In this lesson, practice writing the numerals from 0 to 9. If your student is not ready for this use of his fine motor skills, you may continue by doing the lesson verbally and sharpen counting skills while you write in the correct answers.

Another exercise that may help familiarize your student with his numerals is to use 3 × 5 note cards, or something similar. Write a number on each card and put the correct number of squares on each one corresponding to the number. Color them if you wish.

Then, when the numbers have been learned, have the student put them in sequential order. Encourage the student to say the numbers aloud as he or she plays with the cards.

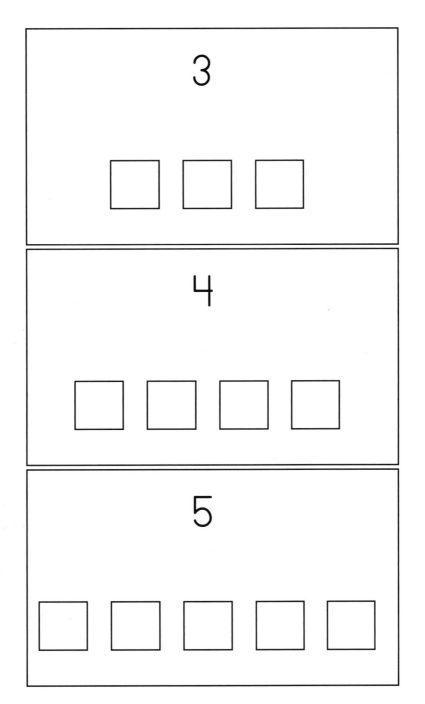

Number Recognition and Writing Numerals

Now we are given the numeral and are asked to color or shade the correct number of squares. The shaded squares don't have to be in any particular spot on the grid. We are focusing on *how many*, not *where*. Study the examples to see how to do these. Place the green unit blocks in the squares before shading or coloring.

Continue to practice writing by tracing over the outlines of the numerals.

Example 1

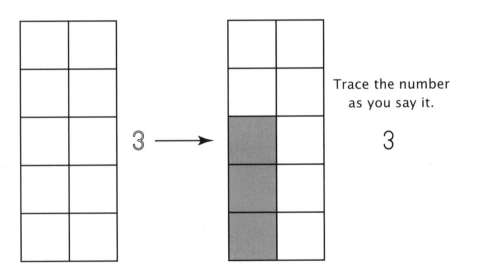

Trace the number as you say it.

3

Example 2

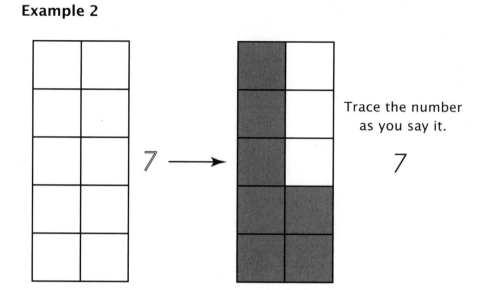

Trace the number
as you say it.

7

A good place to reinforce number recognition and counting is on a *calendar*. During the course of the day, point out the number of the date and notice the date before and the date following.

1	2	3	4	5	6	7
8	9					

At this level, we are focusing on counting from zero to nine. Later, we will add greater numbers. Whenever the student is ready, mention that there are seven days in a week. We will study the different numbers of days in the months when we come to ordinal numbers.

1	2	3	4	5	6	7
8	9	10	11	12	13	14
15	16	17	18	19	20	21
22	23	24	25	26	27	28

Geometric Shapes: Rectangles

A *rectangle* is another way of saying "right angle." ("Rect" is related to a German word that means right.) At this age, we define a rectangle as a four-sided shape with four square corners. Use the rectangles shown to reinforce counting and familiarize the student with this new shape. Have the student count, write, and say the correct answer as you ask the questions.

1. How many rectangles have stripes? (5)
2. How many rectangles are black? (2)
3. How many rectangles are white? (5)
4. How many gray rectangles are shown? (4)

Look around to discover examples of rectangles in your environment.

Number Recognition and Writing Numerals

These lessons are the opposite of lesson 3. The shaded squares are given, and the student is being asked to write the correct numeral. If the writing skills are not developed enough for this yet, write two possibilities and have the student choose and trace the correct answer. Alternately, you may write the number after he or she tells you the answer.

Ask the student to place the correct number of green unit blocks over the shaded squares, count the blocks, and say the number as it is being written.

Example 1

Example 2

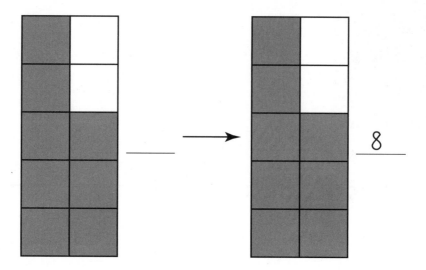

Comparison of Amount

An important math skill that can be taught in the context of everyday life is comparison of amount. It involves ideas such as *more* and *less*, *bigger* and *smaller*, *longer* and *shorter*, and *lighter* and *heavier*. Get in the habit of using these words and others like them whenever appropriate.

You can use the blocks to demonstrate longer and shorter. Use packages from your pantry or other objects to demonstrate heavier or lighter. Be sure that the student understands that one object may be described in different ways. For example, a bookcase may be both taller and heavier than a pencil. Look around you for other ways to help your student understand comparisons.

Geometric Shapes: Circles

Introduce *circles*. Look for circles around you and also draw some. Use a coin, the top of a jar, or a plate to construct circles.

1. How many circles have lines? (4) ⊕
2. How many circles are black? (5) ●
3. How many circles are white? (7) ○

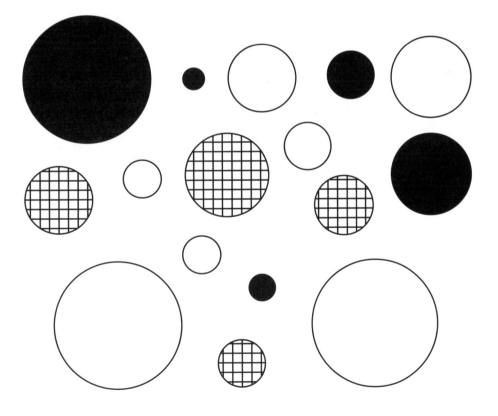

Classifying and Counting

Attribute blocks can be a useful tool for learning the shapes. They are also useful for teaching children that objects can be classified in different ways. Separate the blocks by shape and count how many in each group; then separate them by color and do the same. Ask the student to count red circles or green rectangles.

If you do not have attribute blocks, you can do similar activities with everyday objects. For example, ask the student to separate the red books or the green pencils from a group of objects and count how many of each are present.

Number Recognition and Writing Numerals

This lesson combines the concepts presented in the previous lessons. If the numeral is given, the student reads it and shades the correct number of squares. If the shaded squares are given, the student writes and says the corresponding numeral. Here are some examples.

Example 1

Example 2

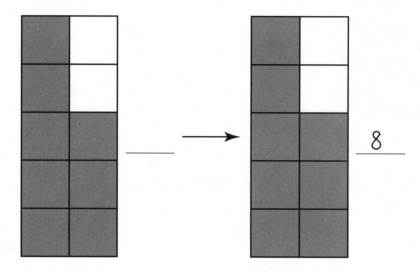

Geometric Shapes: Triangles

Introduce *triangles*. "Tri" means three, and "angle" means angle; a triangle has three sides and three angles. Look for triangles in your environment.

1. How many triangles have stripes? (4)
2. How many triangles are white? (8) △
3. How many triangles are gray? (7) ◮

Teaching Tip

As you look for shapes in the environment, use words that describe their positions. Ask the student if a shape is **above**, **below**, **beside**, **in front of**, **behind**, or **next to** another object. You may also use the Math-U-See blocks to illustrate these different spatial relationships.

Place Value: Units and Tens

I describe this important subject by saying, "Every value has its own place!" To an older child I would add, "Place determines value!" Both are true. There are 10 symbols to tell you how many and many places to represent what kind or what value. Zero through nine tell us how many; *units* and *tens* tell us what kind. For the sake of accuracy, units will be the word used to denote the first value, instead of ones. One is a counting number that tells us how many, and units is a place value that denotes what kind. This will save potential confusion by not saying "10 ones." Remember, one is a number, and units is a place value. The numerals (0, 1, 2, . . . 9) tell us how many tens or how many units. We begin our study focusing on the units and tens, but there are other values, such as hundreds, thousands, millions, billions, and so on.

When teaching this, I like to use a street, since I'm talking about a place. I call the street *Decimal Street*® and have the house for the little green units and the house for the tall blue tens next door to each other. We don't want to forget that, when counting, we only count from zero to nine and then start over. To make this more real, begin by asking, "What is the greatest number of units that can live in this house?" You can get any response to this question from zero to nine, and you might say "yes" to all of them, but remind the student that the greatest number is nine! We imagine how many little green beds or green toothbrushes or green chairs there would be in the house. Ask the student what else there would be nine of. Do the same with the tens. Remember that in the Units house, all the furniture will be green, and in the Tens house it will be blue, like the blue ten bars.

There are directions for making a Decimal Street® poster in lesson 10.

Decimal Street®

Throughout the program, whenever we teach we will employ the following strategy: Build—Write—Say. To teach place value, we will build the number, count how many in each place, write the number, and then read what we've written.

Let's build 42 (four tens, two units). After building, count how many are "at home" in each house. I like to imagine going up to the door of each house and knocking to see how many are at home in each place. Then write the numeral 42 as you count (always beginning with the units) to show the value on paper. Finally, say, "Four tens and two units, or forty-two." Build another number and have the student write how many are at home. When he or she understands this, you write how many on paper and have the student build it! Try 37. After he builds it, you read what he has built. Keep practicing back and forth with the teacher building and the student writing and vice versa.

Here is another exercise I do to reinforce the fact that every value has its own place. I have the student close his or her eyes as I move the pieces around by placing the blue tens where the units should be and vice versa. I then ask the student to make sure the blocks are all in the right place. You might call this "scramble the values" or "walk the blocks home." As the student looks at the problem and begins to work on it, I ask, "Is every value in its own place?"

You've probably noticed the important relationship between language and place value. Consider 42, read as "forty-two." We know that it is four blue ten bars (for-ty, "ty" for ten) and two units. When pronouncing 90, 80, 70, 60, and 40, work on enunciating clearly so that 90 is ninety, not "ninedee." 80 is eighty, not "adee." When you pronounce the number accurately, not only will your spelling improve, but your understanding of place value will also improve. The number 70 (seventy) is seven tens; 60 (sixty) is six tens. The number 40 is pronounced correctly but spelled without the "u" in it. Carrying through on this logic, 50 should be pronounced "fivety" instead of "fifty." The numbers 30 and 20 are similar to 50-- not completely consistent, but close enough so we know what they mean.

The teens can be problematic. Some researchers believe that students in Japan and China have a better understanding of place value than students in Europe and the United States. One of the reasons for this is that, in the Chinese and

Japanese languages, the words for numbers are very regular, and the words for numbers greater than nine are built quite logically from the words for zero to nine. In contrast, there are a number of irregular words for numbers in English and in other European languages, and the English language in particular is very irregular in the words for eleven through nineteen.

To compensate for this, I'm suggesting a new way to read the numbers 10 through 19. You decide whether this method reinforces the place value concept and restores logic and order to the decimal system. Ten is "onety," 11 is "onety-one," 12 is "onety-two," 13 is "onety-three," . . . 19 is "onety-nine." Now it is not that the student can't say "ten," "eleven," "twelve," but learning this method enhances his understanding and makes math logical again. Also, students think it is fun.

When presenting place value, or any other topic in this curriculum, model how you think as you solve the problems. As you work through a problem with the manipulatives, do so verbally so that the student hears your thinking process along with seeing the process. Then record your answer.

Example 1

given visually

As you look at the picture, slowly say it, proceeding from left to right: "fifty-three." Then count, beginning with the units ("1-2-3") and write a "3" in the units place. Then count the tens ("1-2-3-4-5") and write a "5" in the tens place. Do several of these and then give the student the opportunity to do some.

Example 2
74 (given the written number)

Read the number seventy-four, say "seven-ty or seven tens," and pick up seven blue ten bars. Then say "four" and pick up four green unit pieces. Place them in the correct place as you say, "Every value has its own place." Do several of these and then give the student the opportunity to do some.

Example 3
"sixty-five" (given verbally)

Read the number out loud slowly. Then pick up six blue ten bars as you say, "six-ty" or "six tens." Then say "five" and pick up five green unit pieces. Place them in the correct place as you say, "Every value has its own place." Then write the number 65. Do several of these and then give the student the opportunity to do some.

Game for Place Value

Pick a Card - Make up a set of cards with 0 through 9 written in green. Then make another stack of cards written in blue with the same numbers 0 through 9. Shuffle the green cards, pick one, and show that number of green unit blocks. If a child picks a green 4, count out four green unit blocks and show them.

When the child is proficient at this game, try it with the blue cards and do the same as before, except choose blue ten bars instead of the green unit blocks. When the student can do the tens well, add the green cards to the blue cards. Have the child choose one card from the green pile and one card from the blue pile and then pick up the correct number of blue ten bars and green unit blocks.

Place Value: Hundreds

This lesson introduces the *hundreds* place. There is one new house on Decimal Street®, and it is a BIG one. I call it the Hundreds castle. Place value is so important in the base 10 system that we can afford to spend extra time with it. Directions for making Decimal Street® are at the end of the lesson, along with games to reinforce place value. Colored pencils may be easier to use than crayons when doing the worksheets.

If this is too much information or confusing for the student, skip to the next lesson. This topic and all others in *Primer* will be taught again in succeeding books.

Decimal Street®

An important person that I haven't mentioned is Mr. Zero (0). He is a place holder. Let's say you were walking down Decimal Street® and knocking on each door to see who was at home. If you knock on the Units door and three units answer, then you have "3" in the units place. Next door at the Tens house, you knock, and no one answers! This is similar to when a family goes on vacation. You know someone is there feeding the goldfish and taking care of the bird. Mr. Zero won't answer the door because he's not a ten, but he's the one who holds their place until the tens come home. Upon knocking at the Hundreds big red castle, you find that two hundreds answer the door. Your numeral is therefore 203.

Sometime you will want to mention that, even though we begin at the units end of the street and proceed right to left, from the units to the hundreds, when we read the number we do it left to right. We want to get into the habit of counting units first so that when we add, we will add units first, then tens, and then hundreds.

Example 1

given visually

As you look at the picture, slowly say it, proceeding from left to right: "two hundred forty-three." Then count, beginning with the units ("1-2-3") and write a "3" in the units place. Then count the tens ("1-2-3-4") and write a "4" in the tens place. Finally, count the hundreds ("1-2") and write a "2" in the hundreds place. Do several of these and then give the student the opportunity to do some.

Example 2
274 (given the written number)

Read the number "two hundred seventy-four" and then say "two hundreds" as you pick up two red hundred squares. Say "seven-ty" or "seven tens" and pick up seven blue ten bars.

Then say "four" and pick up four green unit pieces. Place them in the correct place as you say, "Every value has its own place." Do several of these and then give the student the opportunity to do some.

Example 3
"one hundred sixty-five" (given verbally)

Read the number out loud slowly. Then, as you pick up one red hundred square, say "one hundred." Then say "six-ty," or "six tens," and pick up six blue ten bars. Then say "five" and pick up five green unit pieces. Place them in the correct place as you say, "Every value has its own place." Then write the number 165. Do several of these and then give the student the opportunity to do some.

Games for Place Value

Pick a Card - This game was described in lesson 9. For this lesson, create one more stack of cards with the numbers 0 through 9 written on them in red. When the child is proficient playing with the green and blue cards, add in the red cards and proceed as before. Place them in three stacks, shuffle, and draw from each stack. Have the student use the blocks to show you the number he has drawn. Example: You select a green 4, blue 7, and red 2. Choose 2 red hundreds, 7 blue tens, and 4 green units. You may want to limit the number of red number cards to correspond to how many hundreds you have. Feel free to make variations. If you have a student who has difficulty writing numbers, use the cards until the fine motor skills are more developed.

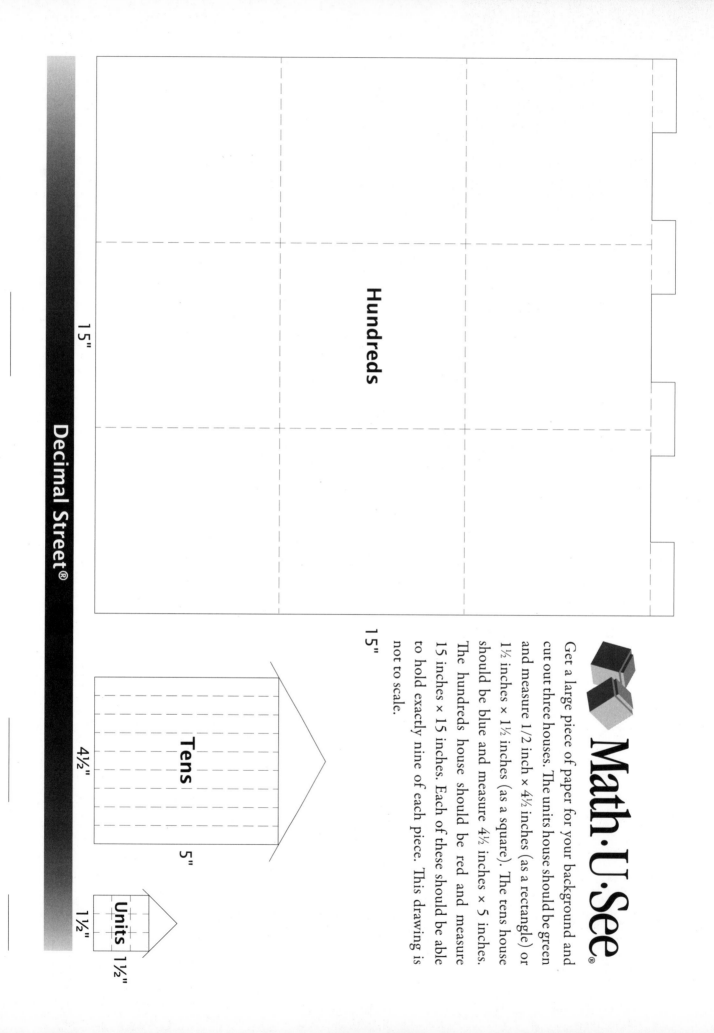

Math·U·See®

Get a large piece of paper for your background and cut out three houses. The units house should be green and measure 1/2 inch × 4½ inches (as a rectangle) or 1½ inches × 1½ inches (as a square). The tens house should be blue and measure 4½ inches × 5 inches. The hundreds house should be red and measure 15 inches × 15 inches. Each of these should be able to hold exactly nine of each piece. This drawing is not to scale.

Hundreds

15"

15"

Tens

4½"

5"

Units

1½"

1½"

Decimal Street®

Unit Bars

Before attempting addition, we want to teach the student that a group of three green units is the same as one pink unit bar, which is three units long. Place the blocks side by side to teach this. Show that if the three individual units are "glued" together, they are the same length as one pink unit bar. This is very important, as this is what addition is: two units plus three units being the same as five units.

In terms of the blocks, transition the student from using the green units exclusively to using the colored unit bars as well. On the worksheets, match the green ones on the left with the correct unit bar on the right. You may also have him or her color the bars on the right to correspond to the green unit pieces on the left.

There are four levels of arithmetic.

1. Counting

2. Adding, which is fast counting

3. Multiplying, which is fast adding

4. Exponents, which are fast multiplying

To move from counting to adding, students have to learn the unit bars so that they can add two plus three and get five, instead of counting "1–2" and then "3–4–5.

Word	#	Unit pieces	Unit bar and color
two =	2 =	▢▢	= ▢▢ orange
three =	3 =	▢▢▢	= ▢▢▢ pink
four =	4 =	▢▢▢▢	= ▢▢▢▢ yellow
five =	5 =	▢▢▢▢▢	= ▢▢▢▢▢ light blue
six =	6 =	▢▢▢▢▢▢	= ▢▢▢▢▢▢ violet or purple
seven =	7 =	▢▢▢▢▢▢▢	= ▢▢▢▢▢▢▢ tan or vanilla
eight =	8 =	▢▢▢▢▢▢▢▢	= ▢▢▢▢▢▢▢▢ brown
nine =	9 =	▢▢▢▢▢▢▢▢▢	= ▢▢▢▢▢▢▢▢▢ light green

Games for Unit Bar Identification

Simon Says - Examples: "Put a three block on your nose" or "Hide two five blocks in a pocket."

What's Missing? - Start with the one through nine blocks. Ask the student to hide his eyes while you remove one of the blocks. Take turns. The variations to this are numerous. Try removing two blocks at a time or starting with two of each number and removing one or two blocks.

The Grab Bag - Put the one through nine blocks in an opaque bag. Take turns either drawing a number card and finding a particular block or telling each other to feel around and find a certain block. The easier version is to simply name the one you are about to pull out. A harder version would be to name the missing block after one has been removed.

Block and Numeral Match Up - Make a set of 3 × 5 cards with the numerals 1 through 9 on them. On the back of the cards, draw colored dots to match the blocks. Place a card with the numeral side showing and match the blocks to the numeral. This is especially good as an activity that students can get out on their own because it is self-correcting. Go ahead and let them peek; they stop soon enough. Introduce the numerals by doing the first three, then the next three, and finally all nine.

Sing and Grab - "If you're happy and you know it . . ." clap three times, grab a five, etc.

Act It Out - Let the student choose a way to act out the number represented by a block. For example, for a three block, clap three times, jump three times, write three lines on a piece of paper, play three notes on the piano, or eat three raisins. Encourage students to use their imaginations for this!

On the worksheets, match the correct answer on the right with the problem on the left.

Example 1

Use the blocks. Count, match, and color. Write the number and say it.

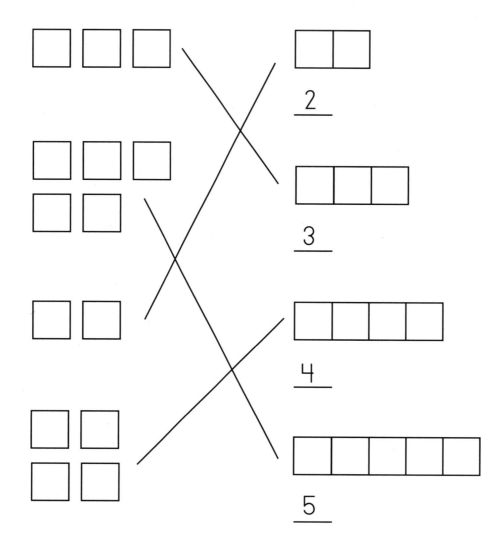

Addition: Introduction and Symbol

As we begin *addition*, think of it as "smooshing" two blocks together and then finding out which block is the same length as the result. If we "smoosh" a five bar and a three bar together, the result is the same length as an eight bar. We build the problem first and then write the symbols: 5 + 3 = 8. When learning the *plus sign*, which is the symbol for addition, maybe a train would help to illustrate the concept. When we take our "5 car" and hook it with the "3 car," the plus sign is the symbol for the connector, and the result is an "8 car." To make it real to a student, perhaps you could say, "If you have five dogs now, and the neighbor gives you three more dogs, how many dogs do you have?" The answer is eight.

Even though the picture is sufficient in many cases, the more you use the blocks, the better will be the understanding on the part of the student.

In the student book, color the unit bars to make the connection from the blocks to the picture. Then write the correct numbers on the lines below the picture and read the equation out loud.

Example 1
Solve: 5 + 3 =

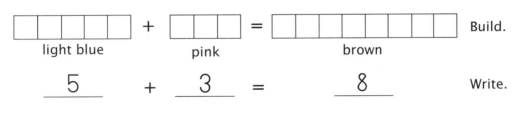

light blue		pink		brown	Build.
5	+	3	=	8	Write.

"Five plus three is the same length as eight."   Say.

Place the five bar and the three bar on top of, or beside, the eight bar. Notice that, with the bars arranged this way, the two lines make an *equal sign*.

5 + 3 = 8

Example 2

Solve: 2 + 7 =

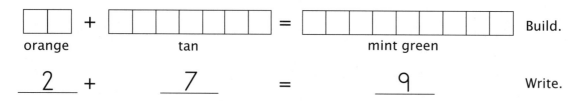

orange   tan   mint green   Build.

_2_ +   _7_   =   _9_   Write.

"Two plus seven is the same length as nine."   Say.

2 + 7 = 9

Addition: +1

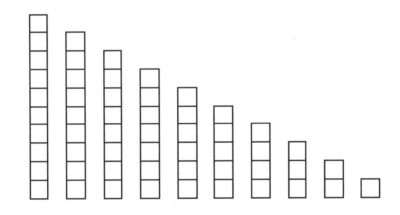

You can use the following as a narrative and adapt it until the concept is learned. Arrange the blocks as shown above. Have the student put his/her finger on the three bar. Say, "Point to the number that is **one more** than this." (four) "What is **one greater** than that?" (five) "If you **add one** to five, what do you have?" (six) "What is **one greater** than six?" (seven) "If I add one to seven, what do I have?" (eight) "What is one more than eight?" (nine) Practice until the student knows "one more than." Change your vocabulary to teach these important words that indicate addition.

When showing the one facts, begin by taking the green unit bar and placing it end to end with the pink three bar. Ask the student if he can find another unit bar that is the same length as the one bar and the three bar "smooshed" together. This is addition: placing a bar end to end with another bar and finding a third bar that is the same length (in this case, the yellow four bar). Beside the bars write 1 + 3 = 4 while saying, "One plus three is the same length as four," or "One plus three equals four." The child sees it, builds it, writes it, reads it, and hears it. After continuous practice, he will eventually understand it and be on his way to remembering it.

Example 1

Solve: 3 + 1 =

Place the 3 bar and the 1 bar
on top of, or beside, the 4 bar.

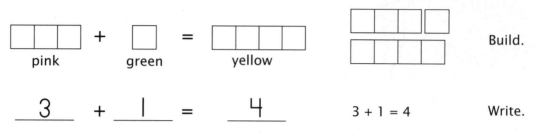

pink   green   yellow   Build.

   3   +   1   =   4   3 + 1 = 4   Write.

"Three plus one is the same length as, or equal to, four."   Say.

Teaching Tip

As you teach the addition facts, help the student make up stories to fit the numbers. For example, "Lily had three grapes. Mom gave her one more grape. How many grapes does Lily have now?" As often as possible, act out the stories that you and your student have made up. Be alert for opportunities to make up number stories that match what you are actually doing in your everyday life.

Counting to 20

The transition from 9 to 10 is pivotal in understanding counting to 100 and regrouping. Work with your student in counting to 20. On the next page there is a story to make counting to 20 more fun and hopefully more understandable.

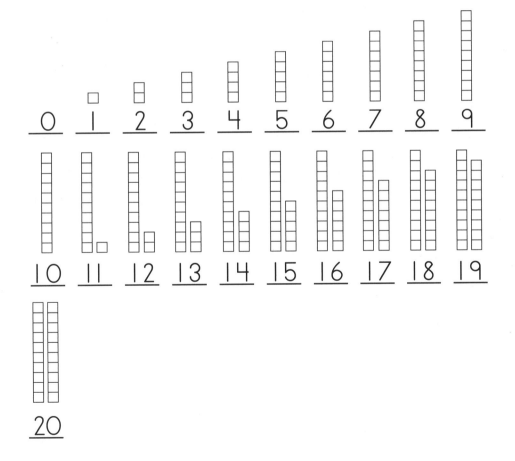

On the worksheets, the students practice writing their numbers on the lines that correspond to the pictures of the blocks. I left a few numbers to make sure your student doesn't miss any.

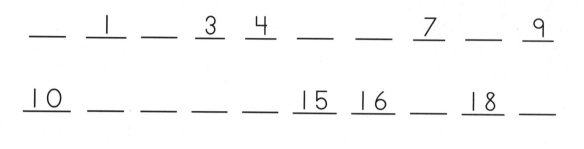

___ | ___ 3 4 ___ ___ 7 ___ 9

10 ___ ___ ___ ___ 15 16 ___ 18 ___

20

Another way to teach counting to 20 uses the first two houses in Decimal Street®. Begin by placing one green unit block in the units house and then say and write "1." Add another unit block and say and write "2." Keep following this procedure until there are nine blocks at home in the units house. Now try to add one more; notice that there is no more room. The reality at this point is that the 10 individual green units are transformed into a blue ten bar.

To make it more interesting, we can make up a story of these new units who keep moving into our neighborhood. Apparently they heard that we have a nice home and naturally want to live with us. After we welcome them into our home, we find that we like them as well. However, when the house is full (nine units) and this new fellow wants to live with us, we have a family discussion. We decide that in order for him to live with us, we will all have to become a blue ten bar and move into the house next door so we can be together. As more units keep arriving, they can stay in the units house because it is now open. When the first unit arrives after we have moved into the tens house, we have one ten bar and one unit bar, or 11. The next unit arrives, and there are 12, or one ten and two units, living on our street. Eventually we have 19 units, and when the next unit arrives, another ten is formed and moves in with the original ten, so now there are two tens or 20. This is illustrated on the next page.

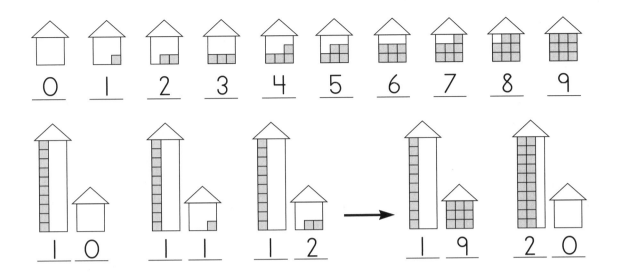

When teaching counting to 20, encourage the student to count aloud using the "nicknames"—one, two, . . . , eight, nine, ten, eleven, twelve, . . . , twenty—as well as the "proper names"—eight, nine, onety, onety-one, onety-two, . . . , two-ty. The nicknames are used most often, but the proper names will help with comprehending place value.

The calendar mentioned in lesson 3 is a good place to apply counting to 20.

Counting to 100

If the student is ready, begin counting up to 100 orally. Use what you have learned about place value to show the numbers with the blocks. Thirty and fifty sound a little different than their "proper names," but the names of the rest of the tens are very logical.

Once the student understands the progression to 100, try building or saying any number under 100 and having the student count on from that point.

Addition: 2 + 2 and 3 + 3; Vertical Addition

There are two new addition facts: 2 + 2 and 3 + 3. Present these with the blocks, building, writing, and saying. When the facts have been learned, one way to reinforce and apply this new skill is by getting dominoes with two and two, as well as three and three, and then asking how many dots are on each face.

You can do the same with two dice with double twos and double threes.

Example 1

Solve: 3 + 3 =

Place the 3 bar and the 3 bar
on top of, or beside, the 6 bar.

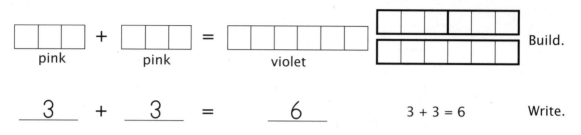

Build.

$$\underline{\quad 3 \quad} + \underline{\quad 3 \quad} = \underline{\quad 6 \quad}$$ 3 + 3 = 6   Write.

"Three plus three is the same length as, or equal to, six."   Say.

Teach the student that you may write an addition problem horizontally and vertically. See Examples 2 and 3. It may be helpful to show the line separating the problem from the answer with two lines and to read it as "equals." Then explain that you really only need one line when doing it this way.

Example 2

Solve: 1
          + 5
       ──────

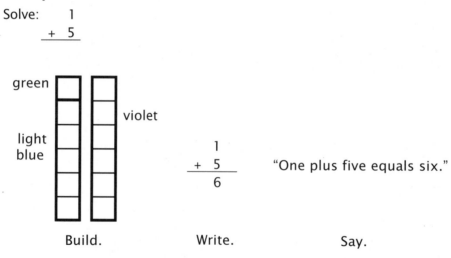

Build.   Write.   Say.

$$\begin{array}{r} 1 \\ + 5 \\ \hline 6 \end{array}$$ "One plus five equals six."

Example 3

Solve: 6
          + 1
          ___

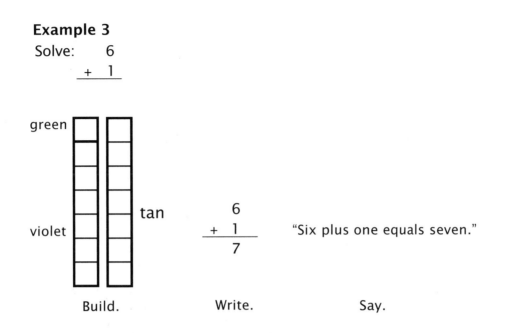

green

violet   tan   6   "Six plus one equals seven."
                           + 1
                           ___
                             7

Build.   Write.   Say.

Teaching Tip

I had a friend who taught at a small private school. Each day, she asked her students to identify the date. Then she would ask them how many ways they could express that number. For example, if the date was 7, it could also be 6 + 1, 5 + 2, or 7 + 0. Using the blocks, the student might also discover that 7 could also be 3 + 3 + 1. Encourage this kind of exploration. The process of breaking down numbers in this way is called ***decomposition*** by educators, but you do not need to use this word with the student.

Shapes: Squares; Addition: 4 + 4 and 5 + 5

A *square* is a special kind of rectangle. It has four right angles (corners) and four straight sides. If the four sides are all the same length, we call the rectangle a square. Tell the student that a square has four right angles (square corners) and four straight sides that are all the same length.

We are going to work on two more addition problems: 4 + 4 and 5 + 5. See Examples 1 and 2 on the next page.

Example 1

Solve: 4 + 4 =

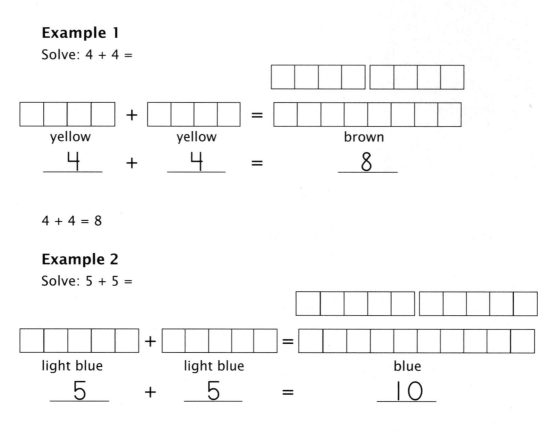

yellow   yellow   brown

  4   +   4   =   8

4 + 4 = 8

Example 2

Solve: 5 + 5 =

light blue   light blue   blue

  5   +   5   =   10

5 + 5 = 10

Teaching Tip

After learning the names of the basic shapes, it is helpful for students to have experience in manipulating the shapes and seeing how they are related to one another. Using the Math-U-See blocks provides some of this practice. For more experience working with different two-dimensional shapes, we recommend attribute blocks or tangrams. Use attribute blocks to sort by color, shape, and size. Tangrams provide challenging practice in putting together different geometric shapes to make new shapes.

LESSON 17

Skip Counting by Two

Skip counting is the ability to count groups of the same number quickly. For example, if you were to skip count by threes, you would skip the one and the two and say "three," skip the four and the five and say "six," and then say "nine, twelve, fifteen, eighteen," etc. If you were to skip count by sevens, you would count "7–14–21–28–35–42–49–56–63–70."

There are several reasons for teaching skip counting.

1. It lays a solid foundation for learning the multiplication facts. We can write 3 + 3 + 3 + 3 as 3 × 4. If a child can skip count, he could say "3–6–9–12." Then he could read 3 × 4 as "three counted four times is 12." As you learn to skip count, you are learning all your multiplication facts in order.

2. Skip counting teaches the concept of multiplication. I had a teacher tell me that her students had successfully memorized their facts but didn't understand the concept. After she had taught them skip counting, they comprehended what they had learned. Multiplication is fast adding of the same number.

3. As a skill in itself, multiple counting is helpful. Skip counting is counting in multiples of a certain number. A pharmacist attending a workshop told me he skip counted when counting pills as they went into the bottles.

4. It teaches the multiples of a number, which are important when making equivalent fractions and finding common denominators. Consider the equivalent fractions 2/5 = 4/10 = 6/15 = 8/20. The numbers 2–4–6–8 are the multiples of two, and 5–10–15–20 are the multiples of five.

In this lesson, we will be counting by twos. When first introducing this, try pointing to each square and, as you count the squares, say the first number quietly; then ask the student to say the second number more loudly. Continue this practice, saying the first number more quietly each time until you are silently pointing to the first square and pointing to the second square as the student says the number loudly. On the homework sheets, only write numbers in the squares with the lines in them. See Examples 1 and 2. In Example 3, we are counting geometric shapes in groups of two.

Some other practical examples of counting by twos are eyes, ears, hands, feet, shoes, and socks. You might ask the student to count all the eyes or shoes in the family or in the classroom. This is also a good time to learn the prefix "bi" that denotes "two" in words such as bicycle or biped.

Another way to teach this skill is with the *Skip Count and Addition Song Book*. Included is a CD with the skip count songs from the twos to the nines.

Example 1
How many boxes are there?

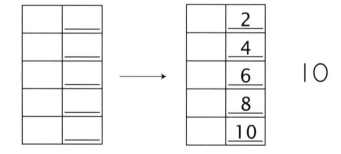

Example 2

How many boxes are there?

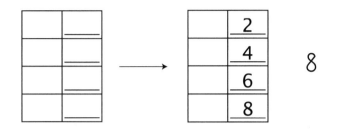

8

Example 3

How many triangles are there?

2-4-6-8-10-12   12 triangles

Addition of Tens

So far all of our adding has been units to units. Two units plus two units is four units. The key phrase to remember here is that "to compare or combine, you must be the same kind." Combining is adding. You can only come up with a new number when you add the same kind. For example, three units plus three units is six units, but three units plus two tens is two tens and three units, or 23.

We have been adding units plus units so far, but now we are going to add tens plus tens.

Example 1

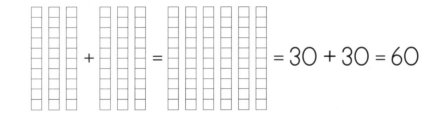

$$= 30 + 30 = 60$$

Verbalized: "Three tens plus three tens equals six tens," or "Thirty plus thirty equals sixty."

Example 2

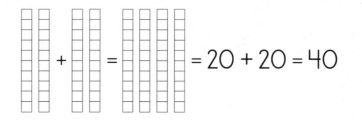

Verbalized: "Two tens plus two tens equals four tens," or "Twenty plus twenty equals forty."

Example 3

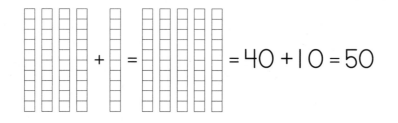

Verbalized: "Four tens plus one ten equals five tens," or "Forty plus ten (onety) equals fifty."

LESSON 19

Skip Counting by 10

After learning to skip count by two, we need to learn skip counting by 10. Some practical examples are fingers on both hands, toes on both feet, and pennies in a dime. Notice that the multiples of ten—twenty, thirty, forty, etc.—all end in "ty." The suffix "-ty" represents ten. Six-ty means six tens, and seventy means seven tens.

Example 1
As with the twos, skip count and write the number on the line. Say it out loud as you count and write. Then write the numbers in the spaces provided beneath the figure.

									10
									20
									30
									40
									50
									60

_____, _____, 30, 40, _____, _____

Example 2

Fill in the missing numbers on the lines.

_____, _____, _____, 40, _____, _____, 70, 80, 90, _____

Solution

10, 20, 30, 40, 50, 60, 70, 80, 90, 100

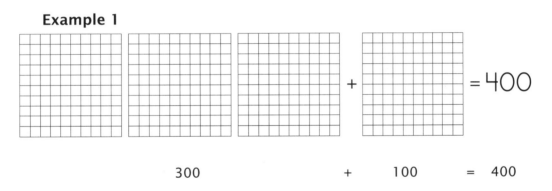

LESSON 20

Addition of Hundreds

Recall that "to compare or combine, you must be the same kind." In this lesson, we apply this principle to adding hundreds to hundreds. When doing this vertically, as in Example 2, make sure your place values are lined up directly beneath one another.

Example 1

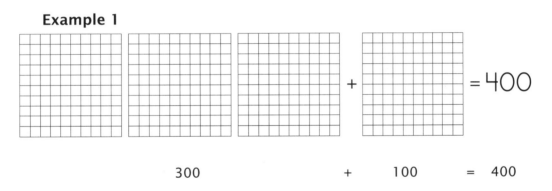

	300		+		100		=	400

Verbalized: "Three hundreds plus one hundred equals four hundreds," or "300 + 100 = 400."

Example 2

$$\begin{array}{r} 300 \\ +100 \\ \hline 400 \end{array}$$

Example 3

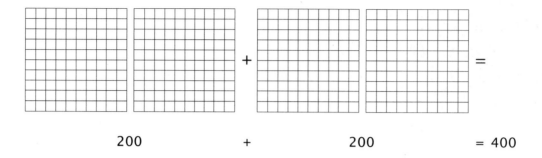

200   +   200   = 400

$$\begin{array}{r} 200 \\ +\ 200 \\ \hline 400 \end{array}$$

Solving for an Unknown

There are three very important reasons for teaching the skill of solving for the unknown while teaching the addition facts.

1. It reinforces the basic facts.

2. It provides a foundation for subtraction.

3. It familiarizes the students with algebra.

Let's do some examples. Notice that we are not teaching algebra abstractly (just letters and numbers on paper) but concretely with the manipulatives to give meaning to the letters and numbers. **When solving for the unknown, allow the student to use the blocks for as long as they are needed.**

Example 1

$\underline{} + 3 = 9$

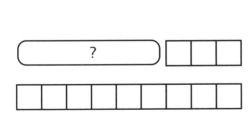

1. Verbalize: "What number plus three is the same as nine?"

2. Place the three bar above the nine bar and find the missing piece.

3. Write six in the blank in the equation.

4. Verbalize: "Six plus three is the same as nine."

The way you verbalize this is essential to understanding the concept. When you are looking for the unit bar, don't be afraid to experiment. I usually reach for the four bar or the five bar first and try it. I want to show that it is okay to eliminate possibilities, and I like to encourage a student's attempts to experiment until he or she finds the correct solution. Later, we can move to $X + 3 = 9$. In that case, we are using a letter because we don't know exactly what number makes it the same, or equal, or an equation. I describe *algebra* by saying, "When we don't know what number to use, pick a letter!"

(Note: Some *Primer* students may not be ready for this next part. Use your judgment.) Although we are happy that the student can solve for an unknown, we are not satisfied until they can make a word problem out of this equation. After all, the end product of our math instruction is to apply it to real-life situations. For our problem, you might say that you need nine dollars by the weekend, and you have three dollars. How many more do you need? While completing the worksheets, encourage the student to make "word problems" to fit the equations. This will help us later when we move in the opposite direction and are asked to write an equation for a word problem.

Example 2

$$\underline{\quad ? \quad} + 2 = 5$$

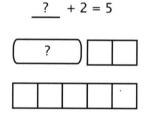

1. Verbalize: "What number plus two is the same as five?"

2. Place the two bar above the five bar and find the missing piece.

3. Write three in the blank in the equation.

4. Verbalize: "Three plus two is the same as five."

This lesson is a good place to practice counting from 0 to 100.

Games for Problem Solving

Who Are You? Who Am I? - Example: "Together we are seven; you are a one, so who am I?" When the student has mastered numbers added up to 10, extend the numbers past 10 if he or she is ready.

Both Sides the Same - Get a large piece of white paper and draw a line down the middle. Place a number bar on one side and a smaller bar on the other side, with a missing piece hidden under a bowl or napkin. For example, you place a seven bar on one side, and showing on the opposite side is a three bar and a bowl. Ask what is hidden under the bowl and say, "What plus three is the same as seven?" When the student figures it out, pick up the bowl and reveal the four bar. Start with lesser numbers and progressively place longer bars on the known side.

Skip Counting by Five

In this lesson, we're teaching skip counting by fives. Use the same techniques as before to introduce and teach this important skill. Some practical examples are fingers on one hand, toes on a foot, pennies in a nickel, players on a basketball team, and sides of a pentagon. We can also remind the student that one nickel has the same value as five pennies and apply the skill of counting by fives to find out how many pennies are in several nickels.

Example 1

As with the twos, skip count and write the number on the line. Say it out loud as you count and write. Then write the numbers in the spaces provided beneath the figure.

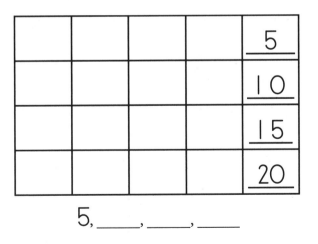

5, _____, _____, _____

Notice that skip counting by five, or fast counting by five, is a way of illustrating multiplication.

The rectangle on the previous page shows five counted four times, or 5 × 4, which is 20. We will use this relationship in later books to help the student see the connection between skip counting and multiplication.

Example 2
Fill in the missing information on the lines.

5, ___, 15, ___, ___, 30, ___, 40, 45, ___

Solution

5, 10, 15, 20, 25, 30, 35, 40, 45, 50

Tally Marks

When you use tally marks, you can only use four lines in a row, with one line for "one," two lines for "two," three lines for "three," and four lines for "four." To show "five," put a slash diagonally through four lines. Tally marks are useful when keeping track of slowly-changing information. Since cowboys used tallies to keep track of their cattle, I think of cows slowly moving through a gate and a cowboy sitting on a fence, making one mark for each cow passing beneath him. Alternately, perhaps you are on a trip, and you want to count red cars that you pass. Each time you pass a red car, you make a line until you get to the fifth one, and then you make a slash.

Look at the chart below to see how to represent the numbers from 1 to 10. Then study the examples as we convert from tally marks to a numeral and then from a numeral to tally marks.

1	I	6	⫲⫲ I
2	II	7	⫲⫲ II
3	III	8	⫲⫲ III
4	IIII	9	⫲⫲ IIII
5	⫲⫲	10	⫲⫲ ⫲⫲

11 ||||| ||||| |

16 ||||| ||||| ||||| |

12 ||||| ||||| ||

17 ||||| ||||| ||||| ||

13 ||||| ||||| |||

18 ||||| ||||| ||||| |||

14 ||||| ||||| ||||

19 ||||| ||||| ||||| ||||

15 ||||| ||||| |||||

20 ||||| ||||| ||||| |||||

Example 1

Make tally marks to show the numeral 7.

7 = 5 + 2 ||||| ||

Example 2

Write a numeral for: ||||| ||||| ||||

||||| ||||| |||| 5 + 5 + 4 = 14

Addition: Making 10

There are two ways to teach addition facts. I call the first the logical way, which is used to teach most of the facts. The second is the "family" method, which we use for the 10 family. Take a ten bar and place it on the table or floor. Using two of the colored unit bars, see how many ways can you make 10. Be sure you verbally say "two unit bars" because addition facts are the combination of two numbers, not three or four numbers. In Figure 1, you can see five ways of making 10. They are 1 + 9, 2 + 8, 3 + 7, 4 + 6, and 5 + 5. Our "10 family" is made up of five different facts. After you build these, write them down and say them.

Figure 1

It is important to teach the facts that make 10 using this "family" method.

Games for Making 10

Build a Wall - See who can build the highest wall that is 10 units long, using two blocks in each row snapped together. After you build your wall, write down the equation that corresponds to the two pieces in each row of the wall. If the first floor is a six bar and a four bar, then the equation is $6 + 4 = 10$.

Fill in the Space - Fill in the blank space with the correct block. You are still building on the ten bar, but instead of placing two blocks to be the same length as the ten bar, you place only one, and the student places the other one. Then it is his turn to place one, and you figure out the length of the missing piece. Example: Begin with a ten bar; then choose a seven bar and snap it on top of the ten bar. Have the student find the piece that makes ten—in this case, the three bar. Have the student write down $7 + 3 = 10$. Then let the student be the teacher and chose a different piece, having you find the missing unit bar and write it down.

Race to a Hundred (and Race You Back) - Use playing cards (face cards removed and aces standing for ones) or the cards you made for Pick a Card. Draw a card, pick up the corresponding bar, and place it on your hundred square. If your next card makes a sum less than 10, just add the block by the original bar. If the card makes a sum greater than 10, put the second bar somewhere else on the hundred square.

On your next turn, pick another number and look for blocks that can even out a line, like a two bar to go with the eight bar. Start at zero and accumulate blocks until you reach 100 and cover the entire hundred square.

Race You Back means you don't have to motivate students to put their blocks away! In this game, you go in reverse and remove the value of the card. When the student gets good at this, require each row of 10 to be filled in from top to bottom as they go. This forces dynamic addition/subtraction to be accomplished. By dynamic, I mean that blocks will have to be "traded in" to get the correct blocks that would fill in the required amount.

Here's an example: the first roll is seven, and second roll is a six. The six would be broken into two threes; one three would fill in the first row along with the seven, and the second three would start the next row.

For the race back to zero, I let them pull the rolled number from anywhere within the hundred. A little dynamic work will be needed at the 100 point and when near zero—go as slowly as necessary. (For example: There is only a five needed to finish, and the student rolls an eight. The eight must be "broken" into a five and a three.)

Example 1

Solve: 3 + 7 =

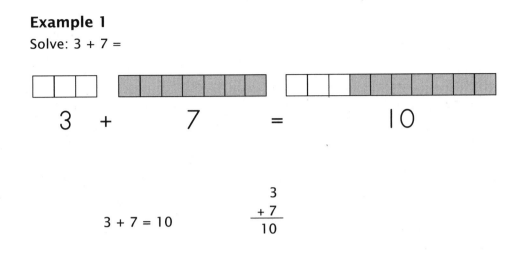

$$3 + 7 = 10$$

$$\begin{array}{r} 3 \\ + 7 \\ \hline 10 \end{array}$$

Example 2

Solve: 6 + 4 =

6 + 4 = 10

$$\begin{array}{r} 6 \\ + 4 \\ \hline 10 \end{array}$$

Example 3

Solve: 8 + 2 =

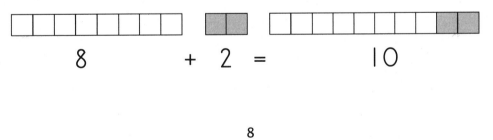

8 + 2 = 10

$$\begin{array}{r} 8 \\ + 2 \\ \hline 10 \end{array}$$

Skip Count to Find Area

In this lesson, the student uses skip counting to find the area of a rectangle. Write the numbers in the spaces where there are lines, as before. Your answer is the last, or greatest, number written.

As you improve at this, you can skip count verbally and then write your answer in the lower right-hand box.

Example 1

Example 2

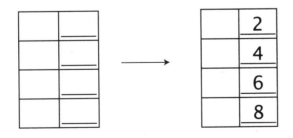

Example 3

									10
									20
									30
									40
									50
									60

Teaching Tip

Students should be able to distinguish between flat (two-dimensional) and solid (three-dimensional) objects. If you have a block set that include cubes, rectangular prisms, cones, and cylinders, you may use them to give students experience in putting these shapes together to make new shapes. At this level, students do not need to use formal names for the solid shapes or to know the words two-dimensional and three-dimensional.

Look for examples of different three-dimensional shapes in your everyday environment, just as you did for two-dimensional shapes. For examples, you could use a box for a cube or rectangular prism, a can from the pantry for a cylinder, and an ice cream cone (of course) for a cone. You can also have the students model shapes with clay.

Telling Time with Minutes

When the student has mastered skip counting by fives, he is ready to learn how to tell time. This can be challenging with a clock that is not digital. We'll begin by taking six of the ten bars and explaining that there are 60 *minutes* in one hour. Next replace each ten bar with two five bars. If you don't have 12 five bars, use five units, or a four bar and one unit, or a three bar and a two bar. Arrange your 12 groups of five in a circle (really a dodecagon, or 12-sided polygon) so you have your 60 minutes in the shape of a clock. Beginning at the top, start skip counting by fives and go around the clock: 5–10–15–20–25–30–35–40–45–50–55–60.

Choose any long unit bar, turn it on its side so that it is smooth, and you have your minute hand. Point it at different areas and, beginning at the top, count your minutes. There is a template to make your clock in the back of the student book. Using all of the blocks in the basic box of blocks, you can make a clock using various combinations for fives (five unit pieces, a two and a three, or a four and a unit).

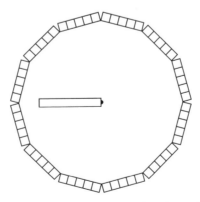

Even though we emphasize skip counting by fives for reading the clock and finding the minutes, you may also remind your student that there are 60 possible minutes, and they are not all multiples of five. When giving problems on your own block clock, move the minute hand to numbers signifying 1, 2, 8, 34, or 51. Reinforce this by having the student count around the clock by ones, saying the multiples of five loudly, as you did when you learned skip counting by fives: *1–2–3–4–**5***.

To help the student see the progression of the minutes, build several partial clocks, as in Examples 1 and 2. Count the minutes by beginning at the top and moving around to the right, or clockwise.

Example 1
Count the minutes: 5-10, or 10 minutes

Example 2
Count the minutes: 5-10-15, or 15 minutes

Telling Time with Hours

When the minutes are mastered, we can add the *hours* by placing a green unit bar at the end of the first five bar (outside the circle) and pointing away from the center of the clock. To distinguish between the minutes and hours, I leave the minutes (unit bars) right side up, but I place the hours upside down with the hollow side showing. You can still see the color and the number of hours, but it helps to distinguish the minutes from the hours. Place your orange bar (upside down so the hollow side is showing) at the end of the second five bar. Continue this process with all the unit bars through 12. Choose a unit bar smaller than your minute hand for an hour hand. Turn this upside down so the student makes the connection between the hour hand and the hours, since both are upside down with the hollow side showing.

Now position the hour hand so that it points between the two and the three. This is the critical point for telling time. Is it two o'clock or three o'clock? I've explained this to many children with success by aiming the hour hand at the two and saying, "He just had his second birthday and is now two." Then I move the hand towards the three a little and ask, "How old is he now? Is he still two?" "Yes." Then I move the hand a little further and ask if he's still two. "Yes." I do this until the hand is almost pointing to three and ask the question, "What about the day before his next birthday; how old is he?" "Still two." He is almost three but still two. Practice this skill until the student can confidently identify the hour by moving the hour hand around the clock.

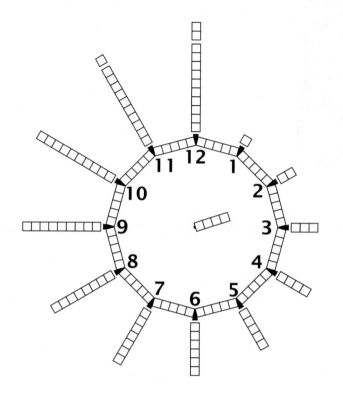

Telling Time with Minutes and Hours

When the hour hand is mastered, put the hours and minutes together. With the hour hand pointing between the three and the four in the example, the hour is three. Skip count by fives beginning at the top to get 30 minutes. Say it as "three thirty" and write it as 3:30.

   When this is mastered, have the student tell time using a real clock.

Example 1

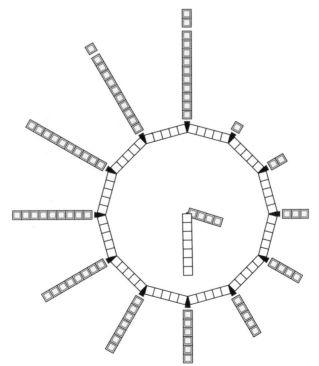

Subtraction: Introduction and Symbol

Subtraction is the opposite, or inverse, of addition. If you know how to add two numbers (2 + 3 = 5) and solve for an unknown (__ + 2 = 5 and __ + 3 = 5), then subtraction will follow easily. Rather than teach another complete set of facts, we will relate subtraction to these two previously-mastered skills. Every subtraction problem can be rewritten and rephrased as an addition problem.

We can read the example 5 – 2 = ? as "What number plus 2 is the same as 5?" This is identical to __ + 2 = 5. You can see this with the blocks in Example 1 on the next page.

The answer to a subtraction problem is called the *difference*. Instead of focusing on taking away or "minusing," focus instead on the difference between the two numbers being subtracted. That is why we can take away, and we can also add up.

Set this up exactly as an addition problem by pushing the blocks end to end. Since subtraction is the opposite of addition, invert the two bar and place it on top of the five bar. The dark blocks in the pictures represent the hollow side of each block. When the hollow side is showing, it means "*take away*", "*minus*", or "owe." One parent commented that you are "in the hole" when the hollow side is showing.

In this lesson, work on rewriting a subtraction problem as an addition problem. Use the blocks to show how subtraction is the inverse of addition. Help the student see how each problem is solved by adding up. Since we are working in *Primer*, use the blocks in every problem. This is an introduction to a subject covered more fully in the next two levels.

Example 1

Solve: 5 − 2 or
$$\begin{array}{r} 5 \\ -\ 2 \\ \hline \end{array}$$

Make this a real problem by saying, "I have five dollars, and I owe the paper carrier two dollars. How many dollars do I have left?"

Step 1

Rephrase as an addition problem.

"Five minus two" is the same as "What plus two is the same as five?"

Step 2

Rewrite as solving for an unknown: ____ + 2 = 5

Step 3

Build as an addition problem.

Step 4

Invert and reposition.

Step 5

Solve and write.

We see that 3 + 2 = 5;
thus 5 − 2 = 3.

$$\begin{array}{r} 5 \\ -\ 2 \\ \hline 3 \end{array}$$

Example 2

Solve: 7 − 3 or

Step 1

Rephrase as an addition problem.

"Seven minus three" means

"What plus three is the same as seven?"

Step 2

Rewrite as solving for an unknown: ___ + 3 = 7

Step 3

Build as an addition problem.

Step 4

Invert and reposition.

Step 5

Solve and write.

We see that 4 + 3 = 7;
thus 7 − 3 = 4.

$$\begin{array}{r} 7 \\ -\ 3 \\ \hline 4 \end{array}$$

Subtraction: −1

As with addition, we'll begin with the simpler facts, such as subtracting by one. See Example 1.

Before teaching subtraction by one, you may want to review counting backward by one with the blocks arranged from one to nine, as we did in lesson 13.

Example 1

Solve 4 − 1 or
$$\begin{array}{r} 4 \\ -\ 1 \\ \hline \end{array}$$

Step 1   Rephrase as an addition problem.

"Four minus one" is the same as
"What plus one is the same as four?"

Step 2   Rewrite as solving for an unknown.

___ + 1 = 4

Step 3
Build as an addition problem.

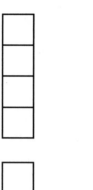

Step 4
Invert and reposition.

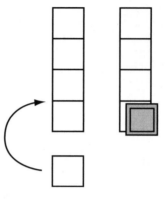

Step 5   Solve and write.

We see that 3 + 1 = 4;   4
thus 4 – 1 = 3. $\underline{- 1}$
                                3

Student Solutions

Lesson Practice 1A

Count the blocks and circle the correct number.

0 1 ②3 4 5 6 7 8 9

0 1 2 3 4 5 ⑥7 8 9

0 1 2 3 ④5 6 7 8 9

Systematic Review 1D

Count and circle the correct number.

0 1 2 3 4 5 6 ⑦8 9

0 1 2 3 ④5 6 7 8 9

0 1 2 3 4 5 ⑥7 8 9

Lesson Practice 2A

LESSON PRACTICE

2A

Cross out the wrong answer. Trace the correct answer.

$\not\!\!X$

8

4

$\not\!\!X$

PRIMER LESSON PRACTICE 2A   19

LESSON PRACTICE 2A

$\not\!\!X$

6

3

$\not\!\!X$

20   PRIMER

Systematic Review 2D

SYSTEMATIC REVIEW

2D

Cross out the wrong answer. Trace the correct answer.

5

$\not\!\!X$

$\not\!\!X$

9

PRIMER SYSTEMATIC REVIEW 2D   25

SYSTEMATIC REVIEW 2D

Count and circle the correct number.

⓪ 1 2 3 4 5 6 7 8 9

0 1 2 3 ④ 5 6 7 8 9

26   PRIMER

Lesson Practice 3A

Systematic Review 3D

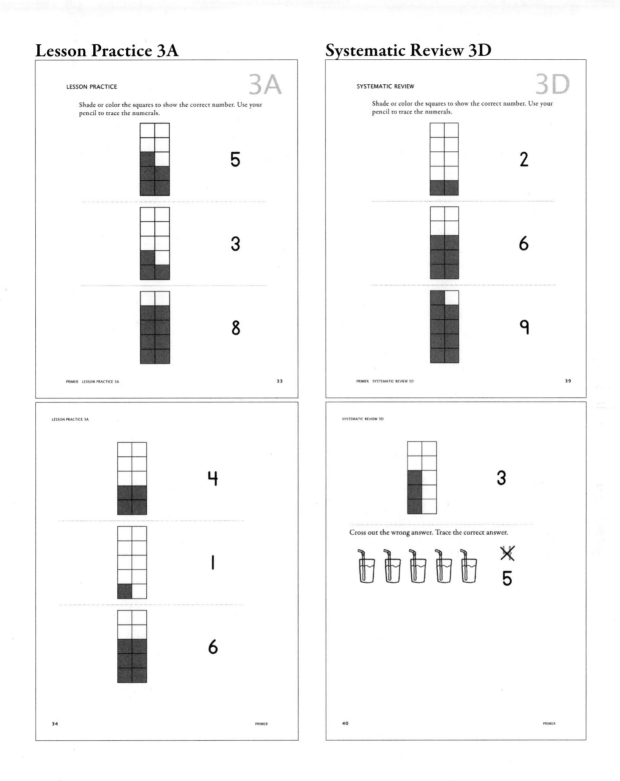

Lesson Practice 4A

Systematic Review 4D

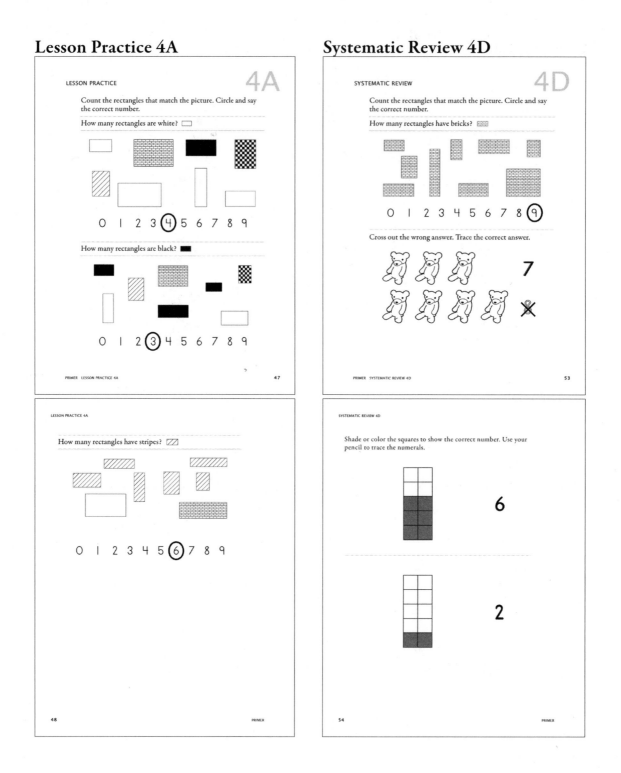

Lesson Practice 5A

Systematic Review 5D

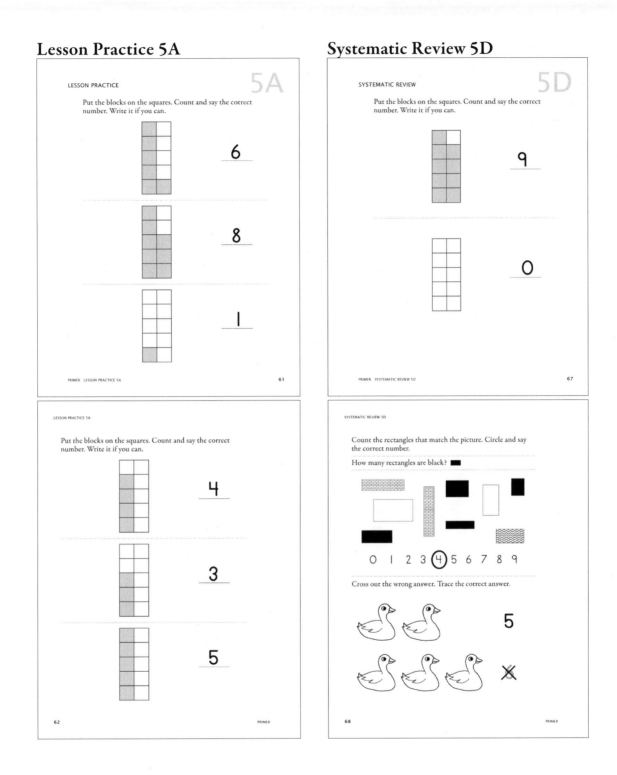

Lesson Practice 6A

Systematic Review 6D

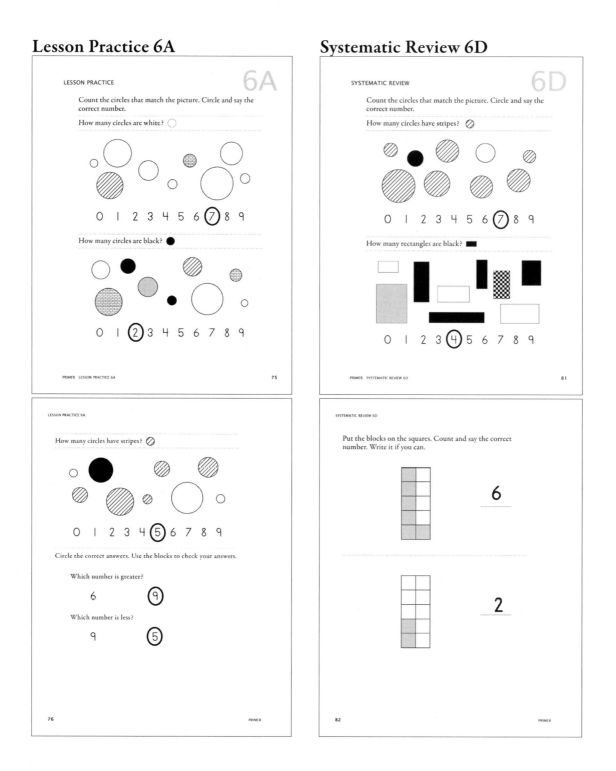

Lesson Practice 7A

Systematic Review 7D

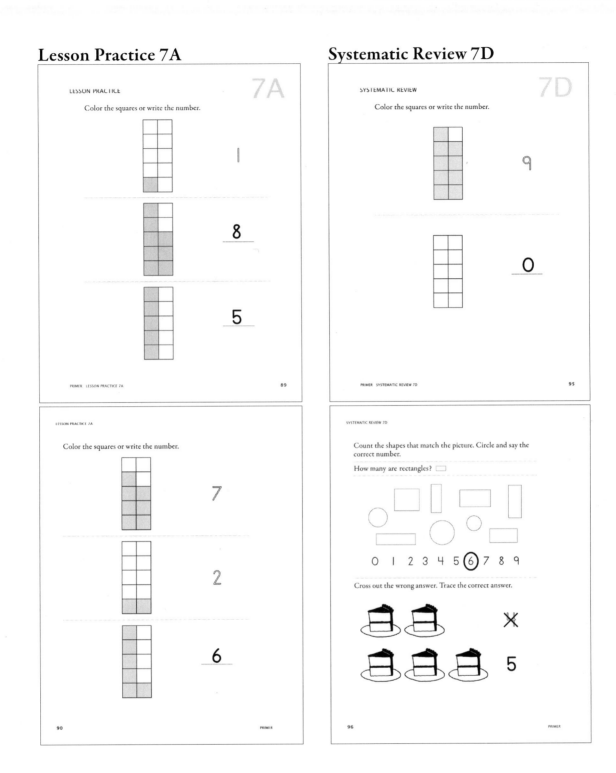

Lesson Practice 8A

Systematic Review 8D

Lesson Practice 9A

Systematic Review 9D

Lesson Practice 10A

Systematic Review 10D

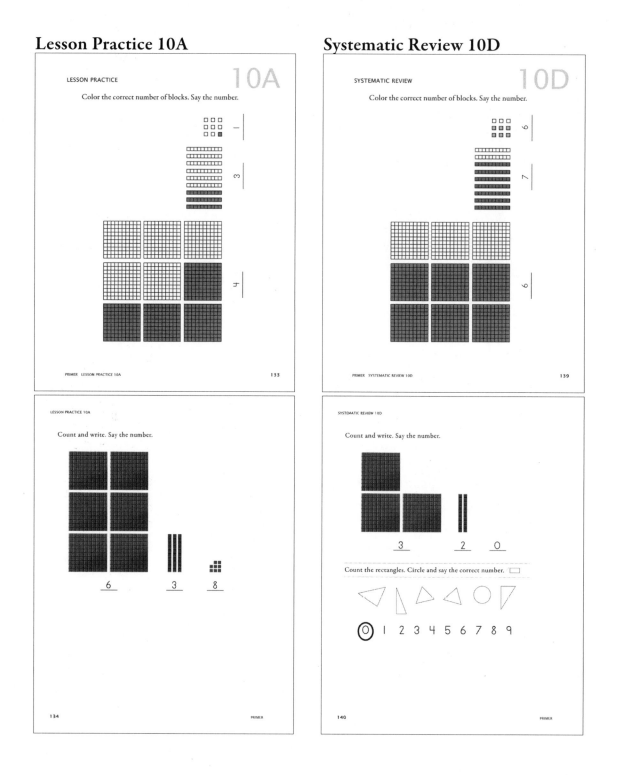

Lesson Practice 11A

Systematic Review 11D

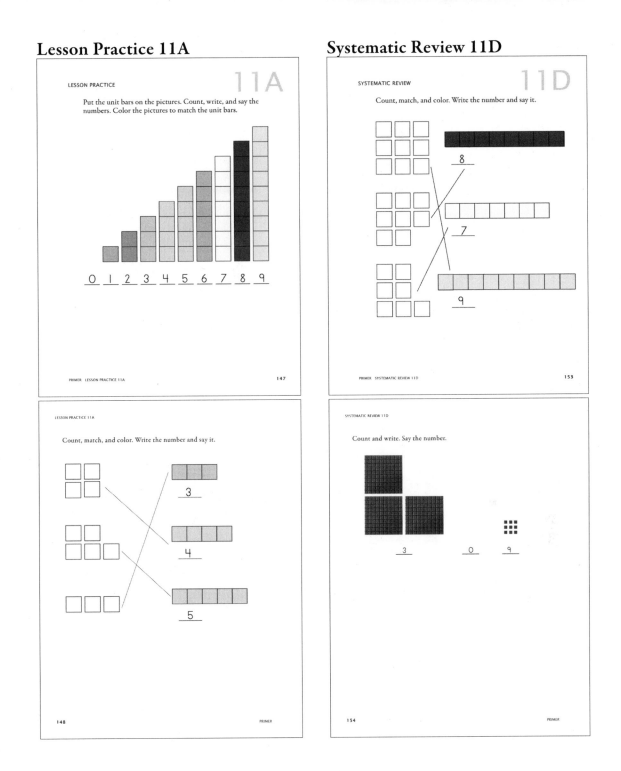

Lesson Practice 12A

Systematic Review 12D

Lesson Practice 13A

Systematic Review 13D

Lesson Practice 14A

Systematic Review 14D

Lesson Practice 15A

Systematic Review 15D

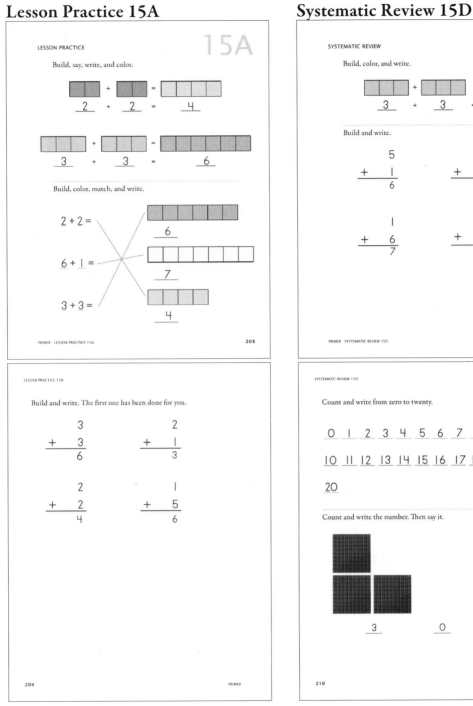

Lesson Practice 15A — 15A

LESSON PRACTICE

Build, say, write, and color.

2 + 2 = 4

3 + 3 = 6

Build, color, match, and write.

2 + 2 =   6

6 + 1 =   7

3 + 3 =   4

PRIMER LESSON PRACTICE 15A   203

LESSON PRACTICE 15A

Build and write. The first one has been done for you.

$$\begin{array}{r} 3 \\ + \ 3 \\ \hline 6 \end{array} \qquad \begin{array}{r} 2 \\ + \ 1 \\ \hline 3 \end{array}$$

$$\begin{array}{r} 2 \\ + \ 2 \\ \hline 4 \end{array} \qquad \begin{array}{r} 1 \\ + \ 5 \\ \hline 6 \end{array}$$

204   PRIMER

Systematic Review 15D — 15D

SYSTEMATIC REVIEW

Build, color, and write.

3 + 3 = 6

Build and write.

$$\begin{array}{r} 5 \\ + \ 1 \\ \hline 6 \end{array} \qquad \begin{array}{r} 2 \\ + \ 2 \\ \hline 4 \end{array}$$

$$\begin{array}{r} 1 \\ + \ 6 \\ \hline 7 \end{array} \qquad \begin{array}{r} 1 \\ + \ 1 \\ \hline 2 \end{array}$$

PRIMER SYSTEMATIC REVIEW 15D   209

SYSTEMATIC REVIEW 15D

Count and write from zero to twenty.

0 1 2 3 4 5 6 7 8 9

10 11 12 13 14 15 16 17 18 19

20

Count and write the number. Then say it.

3   0   6

210   PRIMER

Lesson Practice 16A

Systematic Review 16D

Lesson Practice 17A

Systematic Review 17D

LESSON PRACTICE

17A

How many squares are there? Skip count by two and write the numbers on the lines. Write the number of squares on the line outside the boxes. The first one has been done for you.

	2
	4
	6
	8
	10

10

2	
4	
6	
8	
10	
12	

12

SYSTEMATIC REVIEW

17D

How many squares are there? Skip count by two. Write the numbers on the lines.

2
4
6
8
10
12
14
16
18

18

2
4
6
8
10
12
14
16
18
20

20

LESSON PRACTICE 17A

2
4
6
8
10
12
14

14

2
4
6
8
10
12
14

14

Skip count by two to find how many circles.

○ ○ ○ ○ ○
○ ○ ○ ○ ○

10 circles

SYSTEMATIC REVIEW 17D

Skip count by two to find how many chairs.

16 chairs

Build and write.

$$\begin{array}{r} 7 \\ +\ 1 \\ \hline 8 \end{array}$$

$$\begin{array}{r} 1 \\ +\ 4 \\ \hline 5 \end{array}$$

$$\begin{array}{r} 2 \\ +\ 2 \\ \hline 4 \end{array}$$

$$\begin{array}{r} 1 \\ +\ 1 \\ \hline 2 \end{array}$$

Lesson Practice 18A

Systematic Review 18D

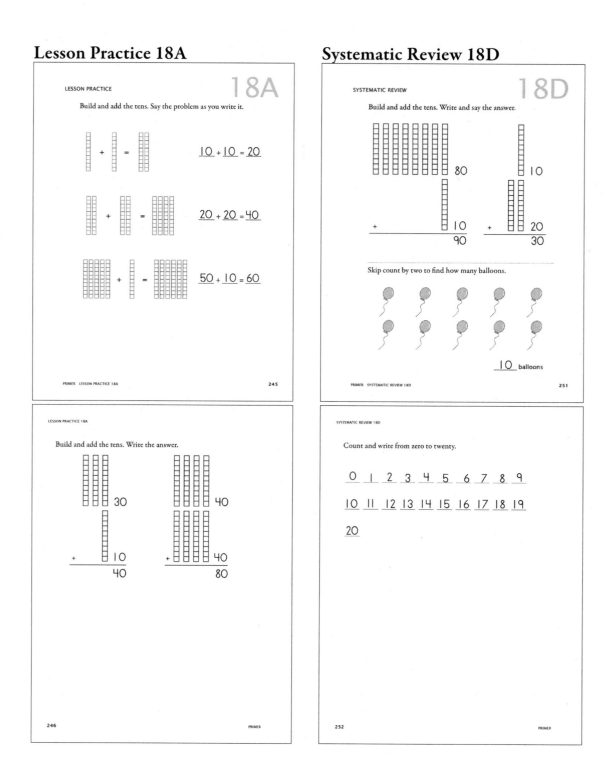

Lesson Practice 19A

Systematic Review 19D

Lesson Practice 20A

Systematic Review 20D

Lesson Practice 21A

Systematic Review 21D

LESSON PRACTICE 21A

Use the blocks to solve for the unknown. Write the answer in the blank and say it.

$\underline{2} + 2 = 4$

$\underline{3} + 3 = 6$

$\underline{6} + 1 = 7$

$\underline{5} + 5 = 10$

PRIMER LESSON PRACTICE 21A   287

LESSON PRACTICE 21A

$\underline{3} + 1 = 4$

$\underline{1} + 4 = 5$

Read the word problem. Use the blocks to help you solve for the unknown number.

Nick wants three toy airplanes. He has two airplanes. How many more airplanes does Nick want?

$\underline{1} + $ ✈✈ $= $ ✈✈✈

288   PRIMER

SYSTEMATIC REVIEW 21D

Use the blocks to solve for the unknown. Write the answer in the blank and say it.

$\underline{4} + 2 = 6$

$\underline{5} + 5 = 10$

Build and write.

$3 + 1 = \underline{4}$

$10 + 60 = \underline{70}$

PRIMER SYSTEMATIC REVIEW 21D   293

SYSTEMATIC REVIEW 21D

Build and write.

$\begin{array}{r} 1 \\ + 8 \\ \hline 9 \end{array}$ $\begin{array}{r} 2 \\ + 2 \\ \hline 4 \end{array}$

$\begin{array}{r} 30 \\ + 10 \\ \hline 40 \end{array}$ $\begin{array}{r} 100 \\ +100 \\ \hline 200 \end{array}$

Skip count by two to find how many triangles.

▽ ▽ ▽ ▽ ▽ ▽ ▽ ▽ ▽
▽ ▽ ▽ ▽ ▽ ▽ ▽ ▽ ▽

$\underline{18}$ triangles

Read the word problem. Use the blocks to help you solve for the unknown.

Jed and Jack have two puppies. One puppy belongs to Jed. How many puppies belong to Jack?

$\underline{1} + $ 🐶 $= $ 🐶🐶

294   PRIMER

Lesson Practice 22A

LESSON PRACTICE 22A

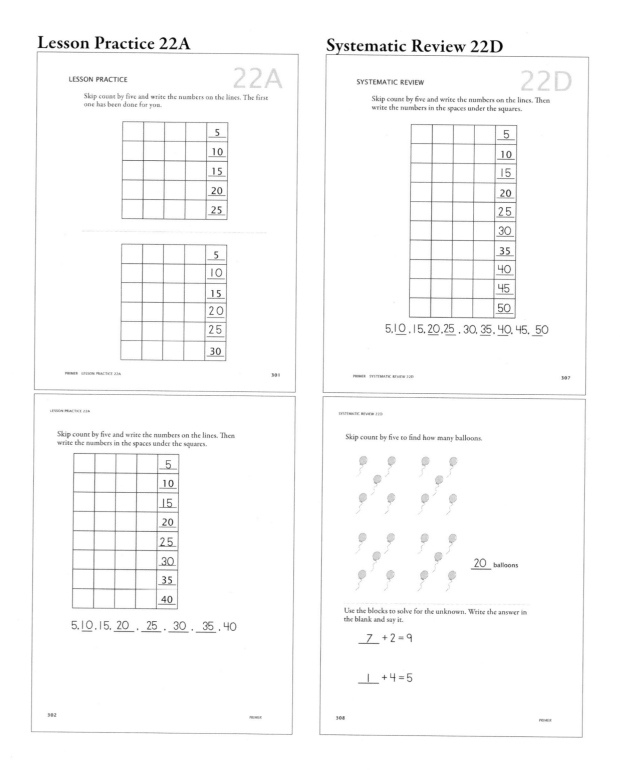

Systematic Review 22D

SYSTEMATIC REVIEW

Skip count by five and write the numbers on the lines. Then
write the numbers in the spaces under the squares.

				5
				10
				15
				20
				25
				30
				35
				40
				45
				50

5, 10 , 15, 20, 25 , 30, 35 , 40, 45, 50

SYSTEMATIC REVIEW 22D

Skip count by five to find how many balloons.

20 balloons

Use the blocks to solve for the unknown. Write the answer in
the blank and say it.

7 + 2 = 9

1 + 4 = 5

116 SOLUTIONS

PRIMER

Lesson Practice 23A

Write the numeral for each set of tally marks. The first one has been done for you.

$\text{卌} = \underline{5}$

$\text{卌 卌} = \underline{10}$

$\text{卌 I} = \underline{6}$

$\text{卌 卌 I} = \underline{11}$

$\text{卌 卌 卌 卌} = \underline{20}$

$\text{卌 卌 卌 卌 卌 I} = \underline{26}$

Use tally marks to write the number. The first one has been done for you.

$17 = \underline{\text{卌 卌 卌 II}}$

$10 = \underline{\text{卌 卌}}$

$6 = \underline{\text{卌 I}}$

$15 = \underline{\text{卌 卌 卌}}$

Systematic Review 23D

Write the numeral.

$\text{卌 卌 卌 I} = \underline{16}$

Use tally marks to write the number.

$10 = \underline{\text{卌 卌}}$

Skip count by five. Write the missing numbers on the lines.

$5, \underline{10}, 15, \underline{20}, \underline{25}, 30, \underline{35}, \underline{40}, \underline{45}, 50$

Use the blocks to solve for the unknown. Write the answer in the blank and say it.

$\underline{1} + 7 = 8$

$\underline{2} + 3 = 5$

Build and write.

$$\begin{array}{r} 6 \\ +\ 1 \\ \hline 7 \end{array} \qquad \begin{array}{r} 4 \\ +\ 4 \\ \hline 8 \end{array}$$

$$\begin{array}{r} 50 \\ +\ 10 \\ \hline 60 \end{array} \qquad \begin{array}{r} 100 \\ +200 \\ \hline 300 \end{array}$$

Lesson Practice 24A

The diagrams are smaller than actual size. Have the student count to select the correct unit bars to complete the problems. Build, say, write, and color.

$1 + 9 = 10$

$3 + 7 = 10$

$2 + 8 = 10$

$4 + 6 = 10$

$5 + 5 = 10$

Build and write.

$$\begin{array}{r} 4 \\ + 6 \\ \hline 10 \end{array} \qquad \begin{array}{r} 1 \\ + 9 \\ \hline 10 \end{array}$$

$$\begin{array}{r} 3 \\ + 7 \\ \hline 10 \end{array} \qquad \begin{array}{r} 2 \\ + 8 \\ \hline 10 \end{array}$$

Systematic Review 24D

Use the blocks to solve for the unknown. Write the answer in the blank and say it.

$\underline{8} + 2 = 10$

?

$\underline{3} + 7 = 10$

?

Write the numeral.

ᴎᴎ ᴎᴎ ᴎᴎ ᴎᴎ ᴎᴎ = $\underline{25}$

Skip count by two to find how many gifts.

$\underline{14}$ gifts

Build and write.

$$\begin{array}{r} 7 \\ + 1 \\ \hline 8 \end{array} \qquad \begin{array}{r} 6 \\ + 4 \\ \hline 10 \end{array}$$

$$\begin{array}{r} 30 \\ + 30 \\ \hline 60 \end{array} \qquad \begin{array}{r} 100 \\ + 100 \\ \hline 200 \end{array}$$

Read the word problem. Use the blocks to help you solve it.

Pam had three dimes. She found three more dimes. How many dimes does Pam have now?

$3 + 3 = \underline{6}$ dimes

Lesson Practice 25A

Systematic Review 25D

Lesson Practice 26A

LESSON PRACTICE 26A

LESSON PRACTICE   26A

Count the minutes. There is a removable clock template near the end of this book.

10   20

15   25

Write how many minutes are shown by each clock.

50   45

Systematic Review 26D

SYSTEMATIC REVIEW   26D

Write how many minutes are shown by each clock.

30   15

Find the area of each rectangle by skip counting. Write your answer in the last box.

4   12   8

Write the numeral.

ΓΗΙ ΓΗΙ ΓΗΙ Ι = __16__

Build and write.

$$\begin{array}{r} 3 \\ +\ 3 \\ \hline 6 \end{array} \qquad \begin{array}{r} 4 \\ +\ 6 \\ \hline 10 \end{array}$$

$$\begin{array}{r} 8 \\ +\ 1 \\ \hline 9 \end{array} \qquad \begin{array}{r} 20 \\ +\ 10 \\ \hline 30 \end{array}$$

Lesson Practice 27A

Systematic Review 27D

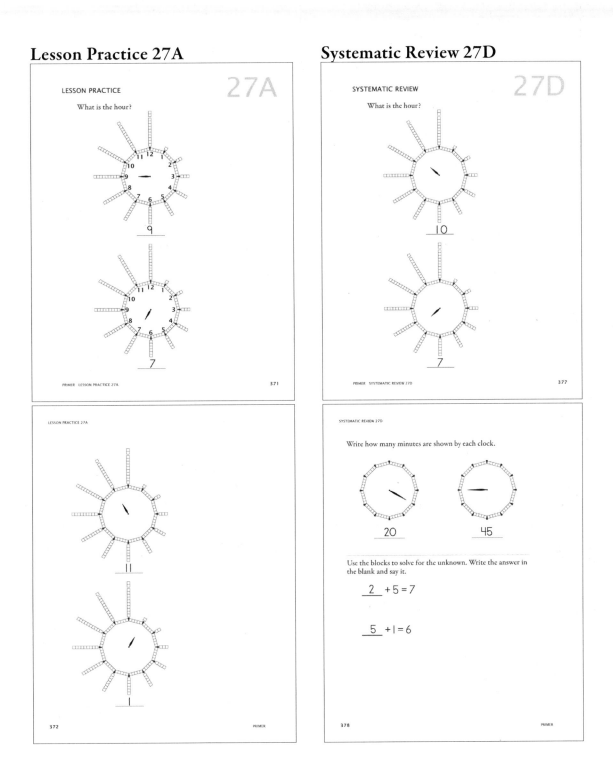

Lesson Practice 28A

Systematic Review 28D

Lesson Practice 29A

LESSON PRACTICE 29A

Build each problem and write the numbers in the circles. The first one has been done for you.

⑥ − ② = ④

⑧ − ③ = ⑤

⑤ − ① = ④

④ − ③ = ①

PRIMER LESSON PRACTICE 29A   399

Build each problem. Say and write the answer.

$$\begin{array}{r} 4 \\ -\ 1 \\ \hline 3 \end{array}$$

$$\begin{array}{r} 5 \\ -\ 3 \\ \hline 2 \end{array}$$

400   PRIMER

Systematic Review 29D

SYSTEMATIC REVIEW 29D

Build each problem. Say and write the answer.

⑤ − ② = ③

② − ① = ①

$$\begin{array}{r} 7 \\ -\ 3 \\ \hline 4 \end{array}$$

$$\begin{array}{r} 6 \\ -\ 1 \\ \hline 5 \end{array}$$

$$\begin{array}{r} 8 \\ -\ 6 \\ \hline 2 \end{array}$$

PRIMER SYSTEMATIC REVIEW 29D   405

SYSTEMATIC REVIEW 29D

Give the time with hours and minutes.

12:30

6:20

406   PRIMER

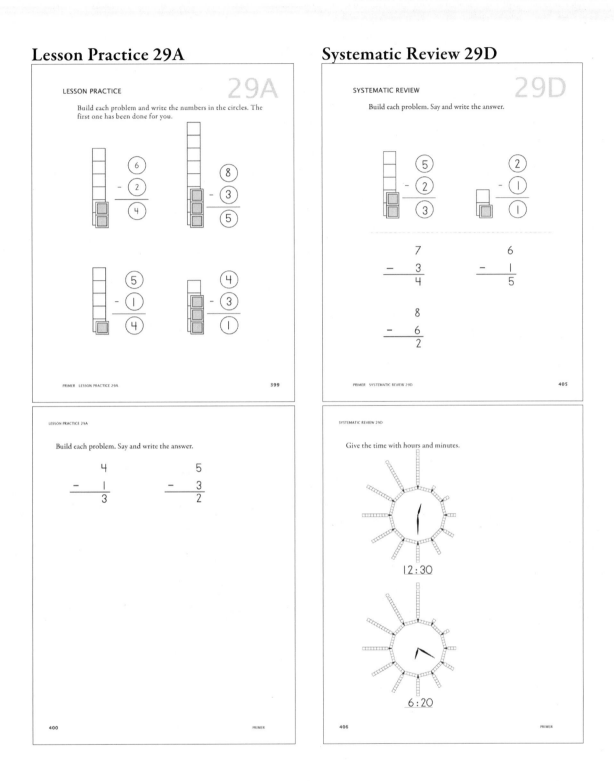

Lesson Practice 30A

LESSON PRACTICE

30A

Build each problem and write the numbers in the circles.

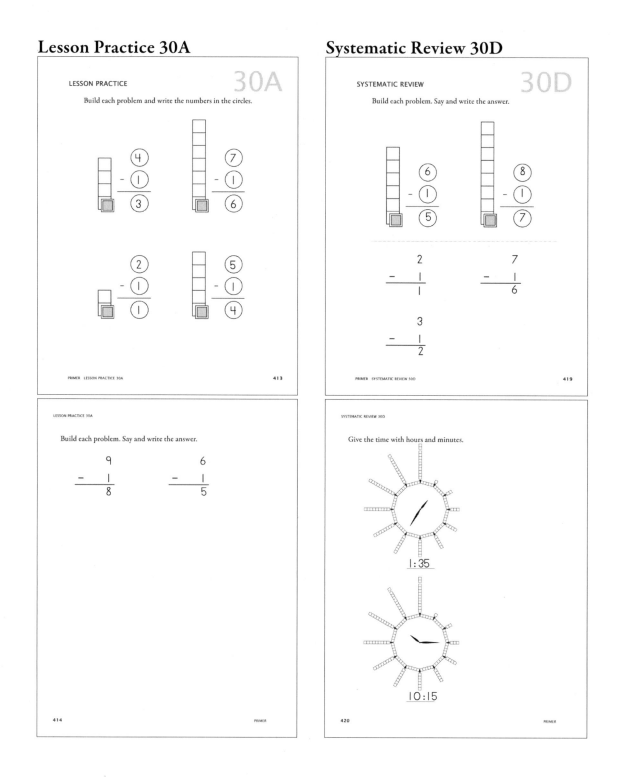

④ - ① / ③

⑦ - ① / ⑥

② - ① / ①

⑤ - ① / ④

PRIMER   LESSON PRACTICE 30A

413

Build each problem. Say and write the answer.

9 - 1 / 8

6 - 1 / 5

LESSON PRACTICE 30A

414

PRIMER

Systematic Review 30D

SYSTEMATIC REVIEW

30D

Build each problem. Say and write the answer.

⑥ - ① / ⑤

⑧ - ① / ⑦

2 - 1 / 1

7 - 1 / 6

3 - 1 / 2

PRIMER   SYSTEMATIC REVIEW 30D

419

Give the time with hours and minutes.

1:35

10:15

SYSTEMATIC REVIEW 30D

420

PRIMER

Master Index for General Math

This index lists the levels at which main topics are presented in the instruction manuals for *Primer* through *Zeta*. For more detail, see the description of each level at MathUSee.com. (Many of these topics are also reviewed in subsequent student books.)

Addition
    facts .... Primer, Alpha
    multiple-digit................................ Beta
Angles .... Zeta
Area
    circle .... Epsilon, Zeta
    parallelogram .... Delta
    rectangle .... Primer, Gamma, Delta
    square .... Primer, Gamma, Delta
    trapezoid.................................... Delta
    triangle...................................... Delta
Average .... Delta
Circle
    area................................. Epsilon, Zeta
    circumference................... Epsilon, Zeta
    recognition .... Primer, Alpha
Circumference .... Epsilon, Zeta
Common factors .... Epsilon
Comparison of amount .... Primer, Beta
Composite numbers...................... Gamma
Congruent .... Zeta
Counting.............................. Primer, Alpha
Decimals
    add and subtract .... Zeta
    change to percent............. Epsilon, Zeta
    divide.. Zeta
    from a fraction .... Epsilon, Zeta
    multiply....................................... Zeta
Divisibility rules .... Epsilon
Division
    facts .... Delta
    multiple-digit.............................. Delta
Estimation .... Beta, Gamma, Delta
Expanded notation.................... Delta, Zeta
Exponential notation........................... Zeta
Exponents.. Zeta
Factors .... Gamma, Delta, Epsilon
Fractions
    add and subtract .... Epsilon
    compare .... Epsilon
    divide .... Epsilon
    equivalent.................. Gamma, Epsilon
    fractional remainders....... Delta, Epsilon

    mixed numbers .... Epsilon
    multiply.................................... Epsilon
    of a number.................. Delta, Epsilon
    of one.......................... Delta, Epsilon
    rational numbers .... Zeta
    simplify .... Epsilon
    to decimals...................... Epsilon, Zeta
    to percents .... Epsilon, Zeta
Geometry
    angles .... Zeta
    area.......................... Primer, Gamma,
    .... Delta, Epsilon, Zeta
    circumference.................. Epsilon, Zeta
    perimeter .... Beta
    plane.. Zeta
    points, lines, rays .... Zeta
    shape recognition .... Primer, Alpha
Graphs
    bar and line .... Beta
    pie... Zeta
Inequalities .... Beta
Linear measure .... Beta, Epsilon
Mean, median, mode.......................... Zeta
Measurement equivalents........ Beta, Gamma
Metric measurement .... Zeta
Mixed numbers................................ Epsilon
Money.................................. Beta, Gamma
Multiplication
    facts .... Gamma
    multiple-digit........................... Gamma
Multiplicative inverse .... Epsilon
Number recognition .... Primer
Number writing............................... Primer
Odd and even numbers .... Alpha, Gamma
Ordinal numbers................................ Beta
Parallel lines .... Delta
Parallelogram.................................... Delta
Percent
    decimal to percent............ Epsilon, Zeta
    fraction to percent............ Epsilon, Zeta
    greater than 100.......................... Zeta
    of a number................................ Zeta
Perimeter.. Beta

Perpendicular lines.............................Delta
Pi....................................... Epsilon, Zeta
Pie graph .... Zeta
Place value...
        Primer, Alpha, Beta, Gamma, Delta, Zeta
Prime factorization..........................Epsilon
Prime numbers .... Gamma
Probability .... Zeta
Rational numbers............................. Zeta
Reciprocal.......................................Epsilon
Rectangle
        area................... Primer, Gamma, Delta
        perimeter .... Beta
        recognition...................... Primer, Alpha
Rectangular solid ....Delta
Regrouping...........................Beta, Gamma
Roman numerals................................Delta
Rounding.....................Beta, Gamma, Delta
Sequencing .... Beta
Skip counting
        2, 5, 10 ....Primer, Alpha, Beta
        all... Gamma
Solve for unknown .... Primer, Alpha,
        ....Gamma, Delta, Epsilon, Zeta
Square
        area................... Primer, Gamma, Delta
        perimeter .... Beta
        recognition...................... Primer, Alpha
Statistics .... Delta, Zeta
Subtraction
        facts .... Alpha
        introduction .... Primer, Alpha
        multiple-digit............................... Beta
Tally marks .... Primer, Beta
Telling time ....Primer, Alpha, Beta
Thermometers and gauges.................. Beta
Trapezoid ....Delta
Triangle
        area.. Delta
        perimeter .... Beta
        recognition...................... Primer, Alpha
Volume of rectangular solid ....Delta

Primer Index

Topic..Lesson

Attribute blocks 6, 16
Addition............ 12, 13, 15, 16, 18, 20, 24
Area.. 25
Calendar 3
Circle 6
Classification 6
Comparison of amount 5
Counting........................ 1, 2, 3, 5, 7, 14
Decimal Street® 9, 10
Decimal system.................................. 1, 9
Decomposition.................................... 15
Games ....
    Act It Out.. 11
    Block and Numeral Match Up 11
    Both Sides the Same 21
    Build a Wall 24
    Fill in the Space 24
    Grab Bag 11
    Pick a Card 9, 10
    Race to a Hundred 24
    Simon Says 11
    Sing and Grab................................ 11
    What's Missing 11
    Who Are You? Who Am I? 21
Hundreds.. 10, 20

Number recognition 1, 3, 5, 7
Place value.................................. 1, 9, 10, 14
Rectangle.. 4
Skip counting
    by 2.. 17
    by 5.. 22
    by 10...................................... 19
    reasons for teaching 17
    to find area.............................. 25
Solving for the unknown 21
Square 16
Subtraction 29, 30
Tally marks 23
Tangrams 16
Teaching tips 8, 13, 15, 16, 25
Telling time 26, 27, 28
Tens 9, 18, 19
Three-dimensional shapes...................... 25
Triangle 8
Unit bars.. 11
Units.. 1, 9, 11
Word problems 13
Writing numerals 2, 3, 5, 7
Zero.. 1, 9, 10